Bacterial Cell Structure

Aspects of Microbiology

Series editors

Dr. J.A. Cole, University of Birmingham
Dr. C.J. Knowles, University of Kent
Dr. D. Schlessinger, Washington University School of Medicine, USA

This series brings important topics of current interest, conceptual difficulty or controversy to the attention of second and third year undergraduates, and to more senior scientists wishing to acquaint themselves with recent developments in fields outside their own specialities. One of the principal objectives of the series is to bridge the gap between introductory texts and research literature. The series editors and the publisher have sought to achieve this objective by providing brief, inexpensive texts which will familiarize students with current research using language and illustrations carefully controlled to conform to undergraduate expectations.

1. Oral Microbiology *P. Marsh*
2. Bacterial Toxins *J. Stephen and R.A. Pietrowski*
3. The Microbial Cell Cycle *C. Edwards*
4. Bacterial Plasmids *K. Hardy*
5. Bacterial Respiration and Photosynthesis *C.W. Jones*
6. Bacterial Cell Structure *H.J. Rogers*
7. Microbial Control of Plant Pests and Diseases *J. Deacon*
8. Methylotrophy and Methanogenesis *P.J. Large*

Aspects of Microbiology 6

Bacterial Cell Structure

Professor Howard J Rogers

Head of Division of Microbiology,
National Institute for Medical Research, London

Van Nostrand Reinhold (UK) Co. Ltd.

First published 1983

ISBN: 0 442 30511 7

Published by Van Nostrand Reinhold (UK) Co. Ltd.
Molly Millars Lane, Wokingham, Berkshire, England

Printed and bound in Hong Kong

Contents

Foreword

The study of bacteria has been a major factor in our understanding of cell metabolism. They are conveniently grown in the laboratory and their rapid rate of growth and multiplication eases the work of the researcher. Many of the chemical structures and reactions that occur in bacterial biochemistry are common to other organisms, or at least have their close counterparts therein, and so it has been possible to use bacteria as the vehicle of a large part of biochemical research. Similarly bacterial genetics, although different from that of eukaryotes in important respects, has much in common with that of other forms of life and consequently bacteria have featured prominently in genetics research. Their rapid multiplication rate is of course a most important feature of such work.

For these reasons the biochemist has used mainly non-pathogenic bacteria for the development of his subject, and their importance in all of this should not be underestimated. In microbiology bacteria have been central to the subject since their first discovery. Moreover, their importance has grown considerably since the discovery of antibiotics. The effectiveness of antibiotics in clinical medicine has done nothing to diminish the need to study bacteria on the grounds that they no longer present a serious challenge to the clinician. On the contrary, the development of resistance towards antibiotics and other chemotherapeutic reagents and the need to understand the details of their mechanism of action have greatly stimulated work on bacterial genetics and biochemistry. More recently the interest in biotechnology and the enormous potential of genetic manipulation have stimulated further the study of bacteria.

Advances in all of these topics require a detailed knowledge of the molecular structure and often the biosynthesis of the components of the cell. Dr Rogers has described in this small book some of the main features of the chemistry and biochemistry of bacterial capsules, walls, membranes, genetic material, flagella and other appendages. Restrictions of space prevent a comprehensive treatment and the author has had to be selective. Further reference is given mainly to reviews so that the reader can discover more about the considerable knowledge that has been gained in this fascinating subject. It is obvious that the biochemistry of bacteria still offers much to the research worker. We know relatively little about the mechanisms of control of biosynthesis and degradation, the translocation of wall and capsular material, the multifunctional nature of such structures and their role in surface interaction and pathogenicity. Much more remains to be discovered about cell growth and division and the possibilities for the alteration of bacterial metabolism through recombinant DNA techniques are still in their early stages of development. It is to be hoped that this book will assist in the stimulation of further research as well as in informing readers of the considerable background that already exists.

James Baddiley

1 Introduction

Ten thousand billion dried bacteria weigh about a gram and a million laid end to end measure only about two metres. Yet they grow, divide, respire, oxidize a truly vast variety of substrates, export toxins, antigens and polysaccharides, and many differentiate. This short book is about the ultrastructure and chemical designs of these small creatures. Considering their size the extent of our knowledge is no small tribute to the ingenuity of microbiologists and to the physicists and engineers who designed and constructed high resolution electron microscopes in the 1930s and 40s. Much of course had already been learnt during the 300 years of developing light microscopy about the shapes and arrangements of micro-organisms long before the advent of the electron microscope. Special staining methods such as that designed by Gram and the acid fast method designed by Ehrlich and Ziehl, both in the nineteenth century, allowed the differentiation of broad classes of micro-organisms. Some cell appendages such as flagella could also be stained but when attempts were made to examine internal structures, the resolution of the optical microscope was strained beyond its proper limits. This led to extensive controversy in many areas, for example, over the presence and behaviour of the nuclear material in bacteria, or over the nature of the walls and membranes.

Is it all worthwhile? The era in which workers tended to look at bacteria as very small bags of enzymes has long past. As soon as more certainty was required about arrangement of enzymes and other macromolecules in the cell morphological understanding became of vital importance. Knowledge of the structure and ultrastructure of larger creatures has made many critical contributions to our understanding of functions of whole organisms. One only has to think of the advances made possible by confirmation of the early 'Cell Theory' to be convinced of the truth of this. If this has been true for the tissues of larger forms of life and of their constituent cells, we would expect it to be equally if not more important for the minute bacteria. When cells from, say, the liver of a mammal are sectioned and examined with an electron microscope their complexity is immediately apparent. We can see mitochondria, themselves the size of bacteria with their own internal structure; systems of internal membranes, some of which bear protein-biosynthesizing ribosomes, and a nucleus wrapped in a complex separate membrane containing the genomic material which separates into chromosomes under the guidance of a spindle apparatus during cell division. Apart from these structures we have the outer bounding membrane, connected with the systems of internal membranes, as well as the membrane bound vesicles, the lysosomes. Bacteria like liver cells, generate energy by oxidizing molecules, pump metabolites from the exterior into the cytoplasm, and biosynthesize proteins as well as a range of other macromolecules. However, they have far fewer internally recognizable structures (Fig. 1.1). Moreover a bacterium, like all cells, must replicate its DNA and separate exactly equal genomes to the two new cells formed each time it divides. All this has to be achieved by cells of about the same size as only one mitochondrion in the liver cells. No well authenticated sub-cellular structures apart from ribosomes have been found in most bacterial species, although some such as the photosynthesizers are exceptions in having

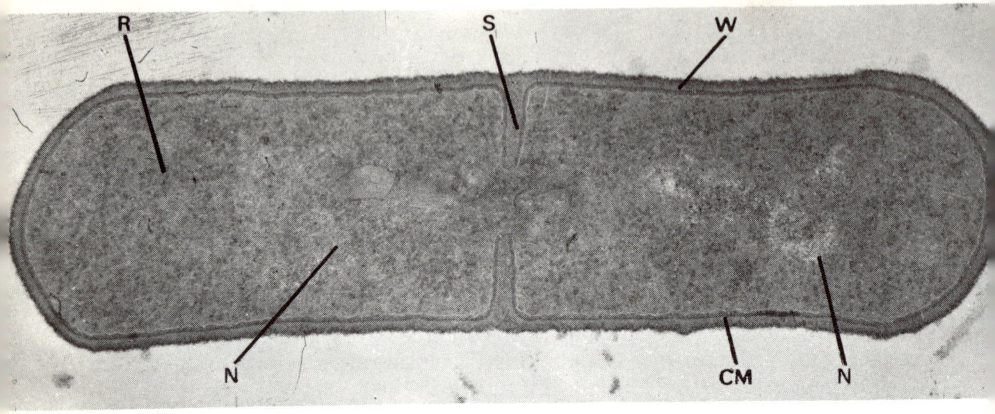

Fig. 1.1 A longitudinal section of *B. subtilis* showing the paucity of recognizable structures within the bacterial cell. W = wall, CM = cytoplasmic membrane, S = septum beginning to divide the cell, N = nuclear material, R = massed ribosomes in the cytoplasm. (Photograph kindly supplied by Dr Ian Burdett)

membranous residues of various shapes; the blue-green algae or cyanobacteria, although true prokaryotes, have an exceptional complexity that differs from both cells of higher organisms and from other bacteria. In some ways they form an intermediate group as far as ultrastructure is concerned. Some others, especially those with well developed specialized functions, such as nitrogen fixation or hydrocarbon utilization, also have relatively simple but plenteous internal membranes. In a number of examples these specialized membranes appear to have been derived by differentiation of the bounding cytoplasmic membrane. Many species also have very simple granules of various sorts packaged in highly specialized membrane-like material.

Unlike the liver cells no neatly packaged nucleus exists in bacteria. Instead of having a number of chromosomes wrapped in the specialized nuclear membrane, the DNA of bacteria exists as a much twisted single circle 'free' in the cytoplasm, often supplemented by smaller circles, the so-called plasmids. Although no specialized membrane surrounds bacterial DNA it is nevertheless probably attached to the cytoplasmic membrane which may form part of the mechanism allowing segregation of the two units of the replicated genome into daughter cells during division.

Space saving has to be effected and multipurpose structures must be used, which is perhaps best illustrated by the many functions undertaken by the simple-looking cytoplasmic membrane. This single structure in bacteria assumes the functions of the mitochondrion as energy generator, the boundary membrane in pumping metabolites in and catabolites out of the cells, the smooth endothelial reticulum and Golgi apparatus in synthesizing polysaccharides for export. It may also possibly serve as the nuclear membrane and spindle apparatus for genome separation. All in all the cytoplasmic membrane is a truly multipurpose organ.

Outside the cytoplasmic membrane is the wall. Terrestrial green plants have evolved thick, strong cell walls that enable them to cope with the force of gravity and to grow away from the earth, whereas bacteria have evolved strong walls presumably to protect the all important cytoplasmic membrane. They have to be able to protect this membrane, not only against noxious physical or chemical agents in the environment but also against a strong internal cell pressure. One of the properties of the cytoplasmic membrane itself is, as has been mentioned, that of pumping a wide variety of low molecular weight metabolites from the external environment, where they may exist at very low concentrations, into the cytoplasm to give high internal concentrations. As a result, an osmotic pressure as large as 20 atmospheres presses on the membrane when some bacteria are surrounded by the usual growth media. This internal pressure must be resisted but at the same time the wall must be able to grow and allow the bacteria to divide. The necessary strength of the wall is nearly always provided by a special class of polymers, the peptidoglycans. Two general plans of wall architecture, both involving the peptidoglycans, have been evolved to provide different sorts of protection in Gram positive and Gram negative bacteria. We now know a good deal about the biosynthetic pathways that give rise to peptidoglycans but have little evidence as to how the walls expand and form septa to divide the bacteria.

Irrespective of whether the outermost layer of the cell is an extension of the bounding membrane as in mammalian cells, or the wall as in bacteria, or capsules which sometimes extend outwards from bacterial cell walls, it is this that makes first contact with the environment, and if very large molecules or even insoluble materials are involved they may be the only points at which interaction can occur. If, for example, bacteria are to adhere to solids, (which they often do in their natural habitats), react with immunogenic systems in host bodies or be able to interact with DNA from another bacterium, they must have surfaces prepared for the particular task in hand. The walls of bacteria are made not only of the necessary strengthening polymers but they also contain a variety of polysaccharides, lipopolysaccharides, proteins and polyol-phosphate compounds distributed to the outermost surfaces. A wide range of polysaccharides, proteins and polypeptides may be excreted to form the semi-organized capsules which extend well out into the environment. It is these components that may decide the nature of the interactions between organisms and their environment. Projecting from some species are a variety of filaments which in one way or another also appear to be involved in cell-cell or cell-environment interactions. Flagella are the propellers for movement and may be singularly attached at one end of the cells, or multiply attached all round them. They can spread bacteria through their environments as well as propel them towards desirable foodstuffs and away from noxious agents. Fimbrae of various sorts, on the other hand, are very thin projecting filaments, some of which are involved in sticking the bacteria to surfaces; somewhat larger filaments in some species form sex pilae that are important both in the transfer of chromosomal DNA from one to another as part of the sexual interactions between the bacteria and in the transfer of plasmids. Plasmids, among other activities, specify resistance to antibiotics. Fig. 1.2 shows a diagrammatic representation of a bacterial cell which summarizes some of the above information.

So far we have spoken of bacterial cells in a general sense. As far as possible this book will accordingly deal with bacterial structures in a general way. The pictures the book contains are of simple shaped rods and cocci. It would be misleading to leave the impression that all bacteria have such simple shapes. There are, for example, those with a spiral shape, the spirochaetes, those with quite complex life cycles each stage having a different shape like the rhodomicrobia or the caulobacters, and the cyanobacteria, which have complex internal structures more similar to eukaryote micro-organisms than to most other bacterial genera. It would require a separate

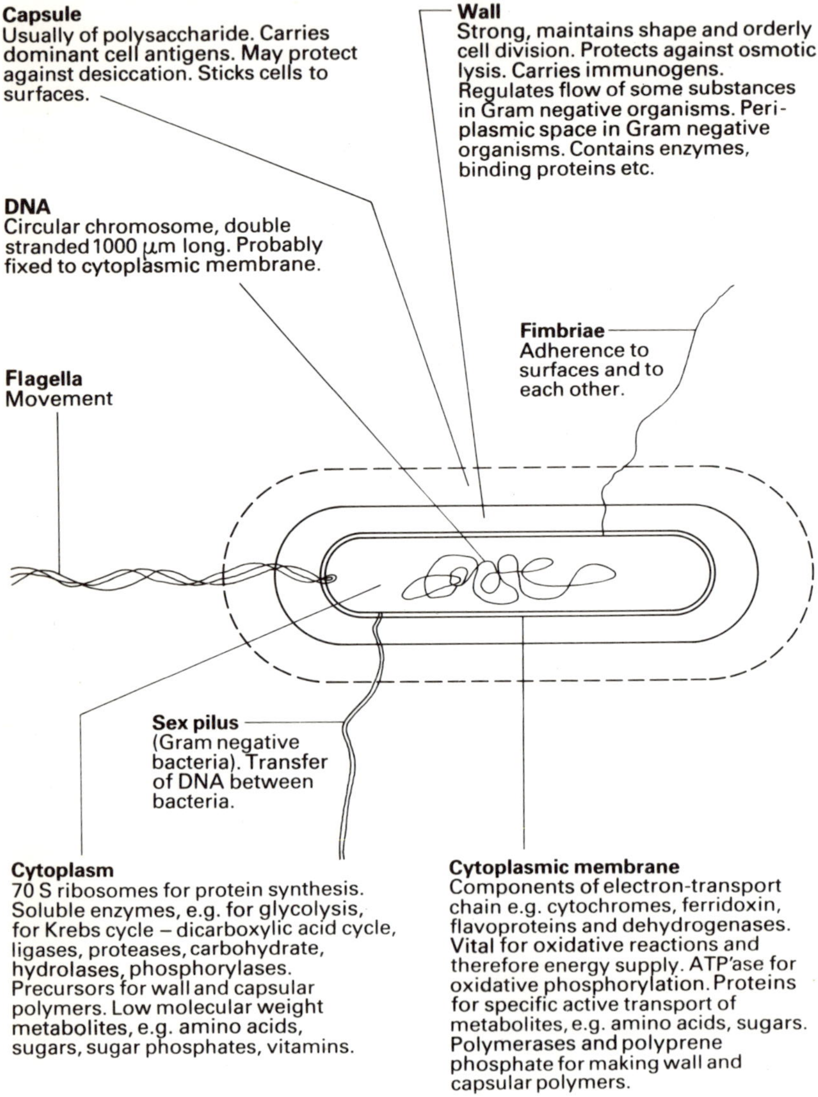

Fig. 1.2 General structure of the bacterial cell.

book however to deal with these more exotic organisms in detail. Nevertheless, as far as is known the basal components of all bacteria are built in the same general way irrespective of their shapes.

Scientific knowledge does not grow evenly within any subject, partly because of the different degrees of difficulty involved in understanding problems and partly because of the different degrees of effort that they have attracted. This frequently leads to paradoxes that may be difficult to understand for those not so familiar with the frequently hard and chancy business of extracting new information from nature. The studies described in this book are by no means free from such paradoxes. We have, for example, extensive and detailed knowledge about the structures of the polymers making up bacterial cell walls but no certain knowledge of how they are arranged. The circular nature of DNA is now well known, as is the arrangement of the genes on it. Yet we know little or nothing about how it is packaged in the cell or how the newly replicated genome is separated from the old. Our understanding of the arrangement of ribosomes within the bacterium is ill defined. When separated from the cell many ribosomes seem to be attached to messenger RNA, which in turn is attached to the DNA. Nothing of this can be seen in the ribosome-packed cytoplasm of the cell. Finally, although we have detailed knowledge about the chemistry of the large number of polysaccharides formed by bacteria as capsules and slimes we have virtually no understanding of the structure of capsules, or of the relationship between capsules and walls. We do not understand the mechanism of decision between the formation of capsules or the liberation of polysaccharide slimes.

Frustrating though these paradoxes may seem, they nevertheless are the very stuff of the excitement and fun of biological research. To watch and contribute to the piecemeal growth of answers to these problems is the joy of those involved in research.

2 The cell walls of bacteria

The appearance of bacterial walls

The specific shapes adopted by bacteria have been recognized from the earliest times of examination by the light microscope. Moreover since rod-shaped bacteria seemed inflexible, that is, when they bumped into objects they did not deform or bend, it was deduced that some sort of stiff layer surrounded them. When placed in strong solutions, the cytoplasm of some bacteria retracted giving further clear indications of a surrounding wall. Attempts, however, to design specific stains for such walls were not very successful, partly because the resolution of the light microscope was taxed too far. This situation was resolved by the advent of electron microscopes and of methods for fixing, staining and cutting sections of bacteria. It then soon became apparent that two basic types of organization of the outer bacterial layers existed and that these corresponded with the Gram staining reaction. This staining method, it will be remembered, consists of first treating a dried and fixed film of bacteria with gentian violet followed by a $KI\text{-}I_2$ solution to form a dark purple complex in the organisms. Subsequent treatment with polar solvents such as alcohol or acetone removes the complex from some species—the Gram negative ones—but not others—the Gram positive ones. Common Gram negative organisms are *Escherichia coli*, salmonellae and gonococci. Common Gram positive species are bacilli, streptococci and staphylococci. Certain exceptions to the general organization of walls exist, and will be dealt with later, but it widely applies and the two types will be described separately. It seems probable that it is indeed the properties of the walls in the two types of organism that lead to the different staining reactions. When walls are removed from bacteria the remaining protoplasts of all organisms stain Gram negative.

Walls of Gram positive bacteria Under the electron microscope, sections of organisms such as bacilli or streptococci from growing cultures, after fixing and staining with osmium tetroxide as described by Ryter and Kellenberg, show a wide, rather transparent layer about 30 nm thick surrounding them. The only infrastructures seen in this layer are two relatively dark bands, one on the innermost side and the other on the outer side (see Fig. 2.1). This difference can still be seen even when no heavy metal stain or fixative has been used to treat the bacteria. The walls dip in to form septa which divide new daughter cells, and outside they are roughened to different degrees. There has been much discussion about the significance of their banded appearance and it was at one time thought to correspond to a true chemical difference in the distribution of polymers. The walls usually consist of about 50 to 60% peptidoglycan, the remainder being made of teichoic and teichuronic acids and polysaccharides. The middle, more electron-transparent layer was thought to be enriched in peptidoglycan. The consensus of modern opinion, however, would be that the layered appearance corresponds to either or both a differing density in an otherwise homogeneous layer or a different penetrability of the stains into the porous electrically charged material.

Although a thickness of 30 nm has been quoted for walls of Gram positive organ-

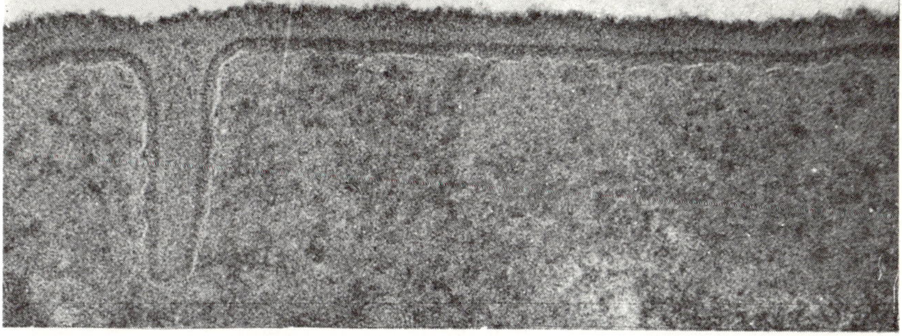

Fig. 2.1 A highly magnified part of a longitudinal section of *B. subtilis* showing the wall in detail with its more dense inner and outer layers. (Photo: Dr Ian Burdett)

Fig. 2.2 The patterned surface layer revealed by the freeze-etching technique on a strain of *Clostridium thermohydrosulfuricum*. (Photograph kindly provided by Dr Sleytr; from Sleytr, 1978)

isms, this can be greatly varied by altering growth conditions. If, for example bacteria are incubated under conditions that allow wall formation but not protein synthesis, uniform thickening over the whole surface occurs. The rate of formation of wall polymers is little affected either by antibiotics which inhibit protein synthesis, such as chloramphenicol or tetracycline, or by omission from media of a required amino acid present in proteins but not in walls. When staphylococci are incubated under these conditions, walls many times thicker than the diameter of the cytoplasm can be formed. The chemical composition of such walls appears to be identical to those around rapidly growing bacteria. Among other consequences of this relaxed control is that in non-steady state cultures, (e.g. older cultures where the bacteria are not growing exponentially), wall thickness may vary greatly.

The walls of bacteria are frequently regarded as the outermost layers of these organisms excluding capsules. The negative staining technique has shown however that this is not always true. In this procedure suspensions of organisms are dried on to electron microscopic grids along with reagents such as ammonium molybdate or sodium silico-tungstate. When specimens are thus treated highly organized patterned layers applied to outer surfaces of both Gram positive and Gram negative species can be seen. The freeze-fracture technique demonstrates these layers even more clearly (see Fig. 2.2). In the latter technique, introduced in the 1950s and perfected in 1961, the sample is frozen as rapidly as possible to a temperature of $-150°C$; this produces only very small ice crystals. It is then transferred to a stage similarly cooled and contained in an apparatus designed so that the sample can be fractured with a sharp glass or diamond knife. More of the structure can be made accessible by subliming some of the ice from the specimen, the freeze-etching process. A thin layer of evaporated carbon or a heavy metal such as platinum is then deposited over the frozen specimen. The resulting replica can be floated off for examination under the electron microscope. The patterned layers seen by either of the above techniques are of protein and are not attached by covalent bonds to the rest of the wall since they can be removed by reagents such as solutions of urea or guanidine hydrobromide that do no more than weaken hydrogen bonding. Table 2.1 shows

Table 2.1 Molecular weights of proteins constituting the patterned layers on the surfaces of bacteria

Micro-organism	Molecular weight (daltons $\times 10^{-3}$)
Acinetobacter strain MJT/F5/199A	67
Clostridium thermohydrosulfurium	140
Clostridium thermosaccharolyticum	140
Bacillus sphaericus (wild type) strain P1	140
Bacillus sphaericus (phage resistant)	
strain MBR 9	105
strains MBR 10–12	86–93
strains MBR, 3, 4, 14, 22, 38, 68	86

the molecular weights of some of the proteins that have been examined. Sheets of reaggregated material can be formed from such protein solutions, and if organisms that have already been stripped of their coats are treated with them together with an appropriate divalent cation such as Mg^{2+} or Ca^{2+}, the original patterned superficial layer can be reformed. The conditions for reattachment and the formation of sheets of the material are rather precise and frequently different. The function of these coats is not clear, although they have some relationship to bacteriophage attachment in *B. sphaericus*. Strains of the organism with different resistance to bacteriophages have coats made up of proteins with molecular weights that differ from those of the parent. However mutants without patterned coats were not found among the resistant strains.

Walls of Gram negative species Unlike the walls of Gram positive species, those of Gram negative organisms are clearly multilayered, with at least two layers that have been isolated and shown to have fundamentally different compositions and organizations. As shown in Fig. 2.3 a thin dense layer overlies the cytoplasmic membrane. This is the rigid or R-layer, and is strength-giving, consisting of peptidoglycan with lipoprotein attached to it. Peptidoglycan accounts for only 5 to 10% of the total dry weight of the wall, unlike Gram positive organisms where the figure is usually 50 to 60% and can be as high as 80 to 90%. The R-layer can be removed by treatment with lysozyme or with other enzymes specifically hydrolyzing peptidoglycan, in the presence of a chelating agent such as EDTA to increase the permeability of the outer membrane.

Fig. 2.3 A highly magnified part of a longitudinal section of *E. coli* showing the multilayered nature of the cell envelope. CM = cytoplasmic membrane, OM = outer membrane, R = rigid layer. (Photo: Dr Ian Burdett.)

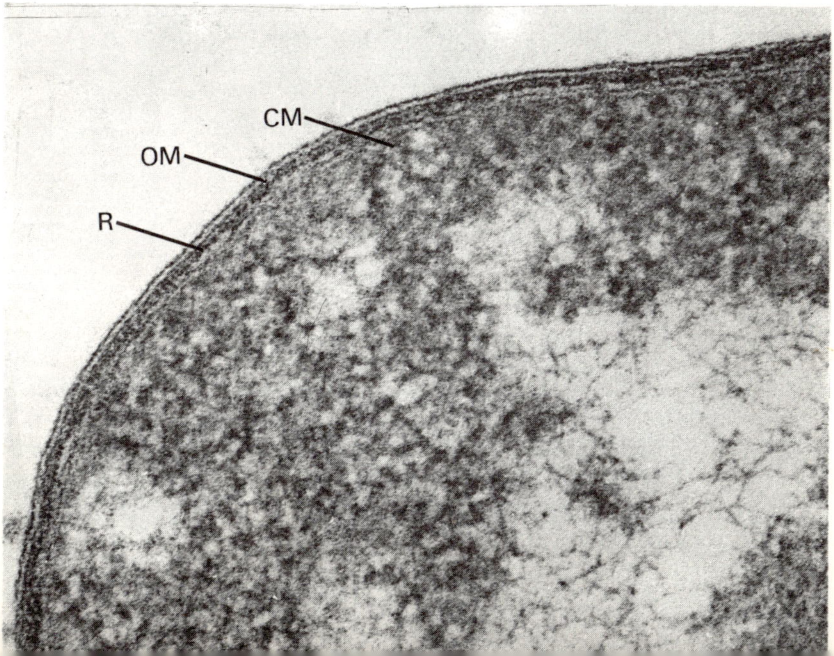

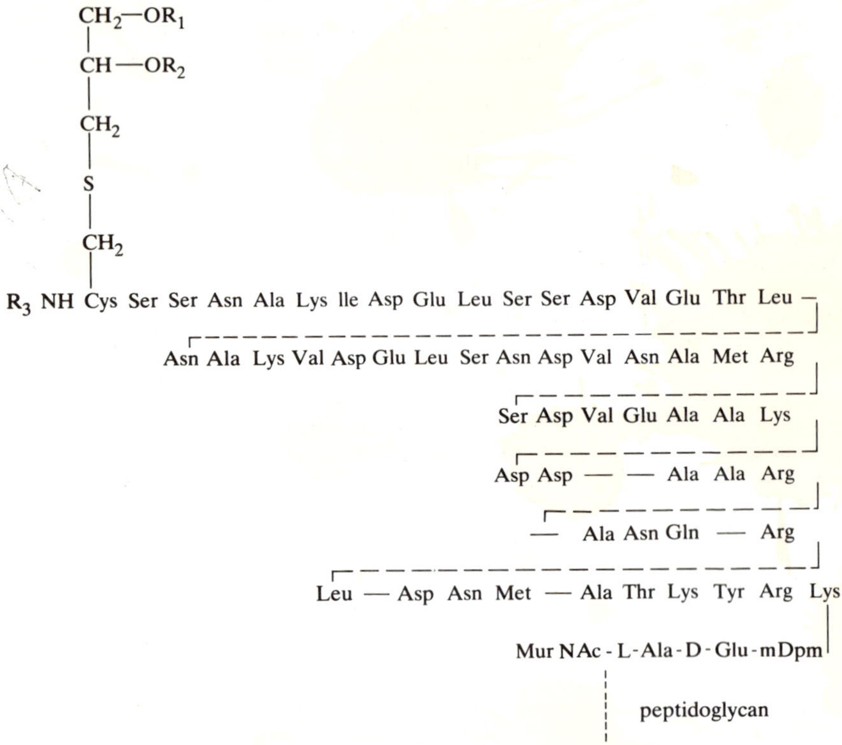

R_1 and R_2 found to consist of 45% palmitic acid, 11% palmitoleic acid, 24% *cis*-vaccenic acid, 12% cyclopropylene hexadecanoic acid and 8% cyclopropylene octadecanoic acid

R_3 is 65% palmitic acid, 11% palmitoleic acid and 11% *cis*-vaccenic acid

Fig. 2.4 The sequence of lipoprotein attached to the peptidoglycan of *Escherichia coli*.

The lipoprotein fraction is covalently attached to the peptidoglycan by the $-NH_2$ of its terminal lysyl residue, whilst at its –COOH end a cystinyl residue is substituted by an esterified glycerol making it strongly hydrophobic (see Fig. 2.4). The lipophilic end of the protein is buried in the outer membrane (Fig. 2.3). It is almost certain that the lipoprotein thus functions as an anchor between the outer membrane and the peptidoglycan layer. Mutants unable to make it show ballooning of the outer membrane away from the rest of the wall.

Lipid A of lipopolysaccharide (see p. 19) is fixed in the outer membrane leaflet and hydrophilic polysaccharide chains extend well out into the surrounding membrane. These chains of the lipopolysaccharides define the immunological specificity of the

O antigens in such organisms as the salmonellae. The outer membrane itself, as we shall see, differs considerably in composition from more usual membranes, not only in its high content of lipid A, but also because of its protein composition which is dominated by the presence of four or five fractions in very much larger quantities than any others. The nature and function of some of these proteins will be discussed in Chapter 3.

If Gram negative bacteria such as *E. coli* are suspended in strong sucrose or salt solutions the protoplast contained by the cytoplasmic membrane shrinks away from the thin peptidoglycan layer. Such a phenomenon is called plasmolysis and can occur in most cells bounded by a wall. It is due to the removal of water from the cytoplasm by osmosis. The osmotic pressure within Gram negative species is, incidentally, much lower than that in Gram positive bacteria although plasmolysis is easier to see in the former. In a considerable number (about 200) of places the layers of the envelope in Gram negative species appear to be anchored together. If plasmolysis is extreme the cytoplasmic membrane is drawn out into long threads attached by one end to the undeformed peptidoglycan layer. These areas of adhesion or junctions may be very important since they correlate topographically with the points on the bacterial surface at which bacteriophages adsorb, lipopolysaccharide is excreted, and at which at least one of the major outer membrane proteins is inserted.

Polymers in bacterial walls

Peptidoglycans The only groups of bacteria that have so far been reported to be without peptidoglycans as supporting polymers in their walls are the archaebacteria, which are so called (see p. 25) to indicate their possible early and separate evolution; among them are the methanogenic bacteria and the halobacteria. The acholeplasma also have no recognizable walls but can perhaps only be regarded as honorary bacteria. Stable L-forms of bacteria are wall-less but are derived directly from normal bacteria, probably by genetic lesions in their wall biosynthetic pathways. Morphologically more complicated organisms such as the cyanobacteria and the actinomycetaceae do have a peptidoglycan layer in their walls.

The peptidoglycans are universally built of *N*-substituted glucosamine and muramic acid (3–0 lactylglucosamine) linked together by $1 \rightarrow 4\beta$ bonds. Very commonly the *N*-substituents are acetyl groups although in mycobacteria, *Nocardia* and *Micromonospora* species glycollyl groups are present and in strains of some species of bacilli high proportions of glucosamine with unsubstituted $-NH_2$ groups have been found. The polysaccharide chains are then linked together by short peptides constituted, in any one peptidoglycan, of a limited number of amino acids (Fig. 2.5). The peptides are linked to the carboxyl groups of the muramyl residues by amide or pseudo-peptide linkages. The amino acids in the peptides alternate between the L and D forms unless there is either a glycyl residue or an amino acid that has two active centres such as 2, 6-diaminopimelic acid. The latter is often found in the *meso* form. Variable proportions of these peptides are linked by their carboxyl termini, always represented by D-alanyl residues, often to amino groups of the L-centre of a diamino acid (e.g. lysine, diaminopimelic acid) in a neighbouring peptide chain which is usually supposed to be substituted on to a separate glycan strand. However, other types of cross-linking are known (see Fig. 2.6). Such peptide cross-linking of the glycan strands is illustrated in Fig. 2.5 and must build up a com-

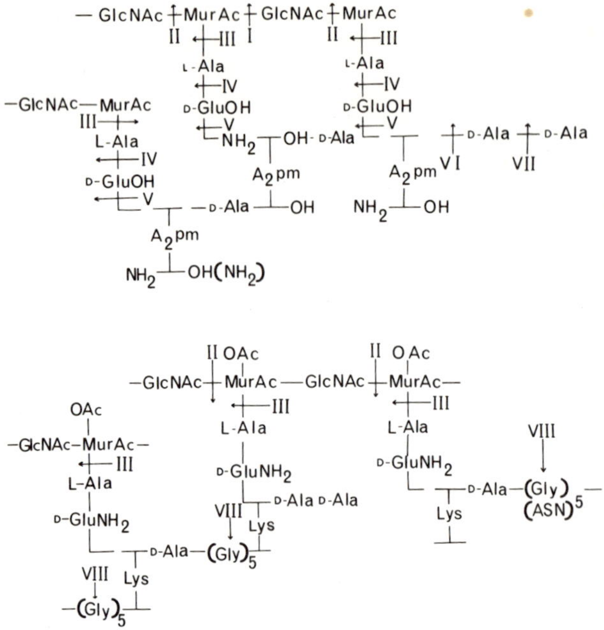

Fig. 2.5 Structure of the peptidoglycans from the walls of (**a**) many bacilli and Gram negative species, (**b**) *Staphylococcus aureus*, in which the crosslinking bridge between peptides consists of five glycyl residues (residues A_4, A_5, A_6 in Figure 2.6), and of *S. faecalis* in which it consists of one asparagine residue. The Roman numerals I, II, III and VIII represent bonds hydrolyzed by lysozyme (I), and naturally occurring autolysins (II, III and VIII) dissolving the walls. VI and VII are the points of attack of D-alanyl carboxypeptidases present in many organisms. IV and V are the points of attack of enzymes found in sporulating cultures.

plicated 'chain armour' around the organisms. Thus the peptidoglycans have a very special structure not easily attacked by enzymes which hydrolyse either common polysaccharides or proteins. Their strength and conformation are influenced by the nature of the cross-linked peptides as well by the $1 \rightarrow 4\beta$ bonds in the glycan. The latter are the same as those in other simple structural polysaccharides such as cellulose (poly $1 \rightarrow 4\beta$ glucose) or chitin (poly $1 \rightarrow 4\beta$-*N*-acetylglucosamine). It should be noted, however, that the substitution of the lactyl group in the three position of the muramyl residues in peptidoglycan prevents the formation of one of the intramolecular hydrogen bonds per disaccharide present in chitin.

Four basic types of peptidoglycan have so far been recognized (Ghuysen, 1968) that are differentiated by the way in which the peptides are linked together. These are illustrated in Fig. 2.6. The range of amino acids found in peptidoglycans from different species of bacteria is, however, large (Table 2.2). A more complicated classification of the different types of peptidoglycan has also been designed (Schleifer and Kandler, 1972). Apart from the variety of amino acids and types of cross-linkage,

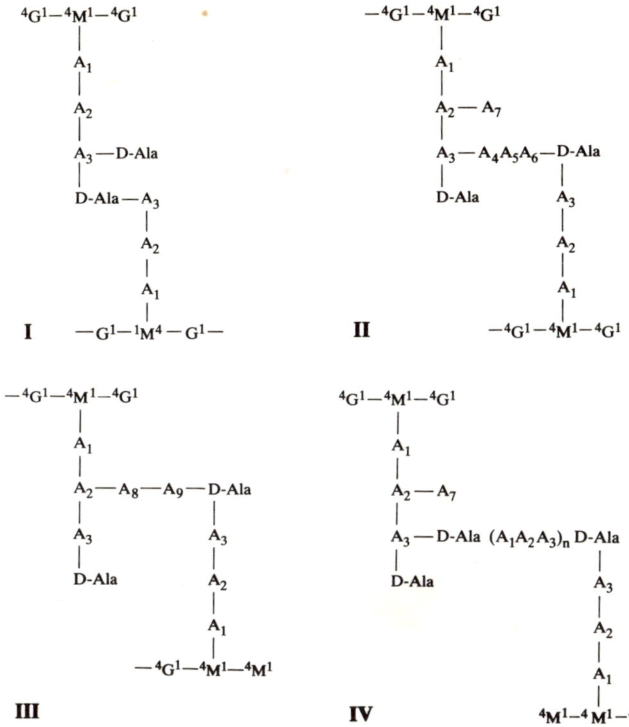

Fig. 2.6 Generalized structures of peptidoglycans G = *N* − acetylglucosamine, M = *N* − acetylmuramic acid, $A_1 - A_9$ are amino acids. A_1 is commonly L-alanine, A_2 is commonly D-iso-glutamate or D-iso-glutamine, A_3 may be a variety of diamino acids although L-lysine and meso-2, 6 diaminopimelic acid are common. A_4, A_5 and A_6 are commonly glycine, serine or threonine (see Table 2.2).

Table 2.2 The amino acids present in all different known peptidoglycans (from Rogers, 1974)

Position	Amino acids that have been identified
A_1	L-Ala(+ +), Gly, L-Ser
A_2	D-iso-Glu(+ +), 3-Hyg
A_3	L-Lys(+ +), meso-DAP(+ +), DD-DAP, L-DAB, L-Hse, LL-DAP L-Ala, L-Glu, L-Orn, meso-HyDAP, L-Hyl, Nγ-acetyl-L-DAB
A_4, A_5, A_6	Gly(+ +), L-Ala(+ +), L-Thr(+ +), L-Ser(+ +), D-Asp(+ +) D-Ser, D-Glu, D-Glu(NH_2), D-Asp(NH_2), L-Lys
A_7	Gly, Gly(NH_2), D-Ala(NH_2)
A_8, A_9	D-Lys, L-Lys, D-Orn, Gly, D-DAB

The positions of A_1 to A_9 are shown in Figure 2.6. The sign (+ +) after an amino acid indicates that it has been found in the peptidoglycan of many different micro-organisms. A_4 to A_6 and A_8 and A_9 are bridge amino acids. In any given organism, the bridge consists of one or two repeating residues, except for certain micrococci, in which three amino acids are present. When D-Asp, D-Asp$(NH)_2$ or D-Glu(NH_2) form the bridge they do so as single residues. Most of these data have been summarized from the review of Schleifer and Kandler. Key: 3-Hyg, threo-3-hydroxy-glutamic acid: DAP, 2, 6-diaminopimelic acid; DAB, 2, 4-diaminobutyric acid; HyDAP, meso-2, 6-diamino-3-hydroxy-β-pimelic acid.

(a) Glycerol teichoic acids

(i) 1, 3 linked compounds

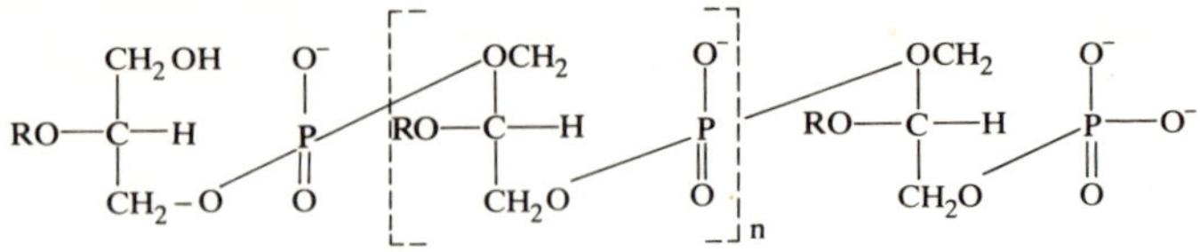

R=H, D-alanyl, or glycosyl

(ii) 1, 2 linked compounds

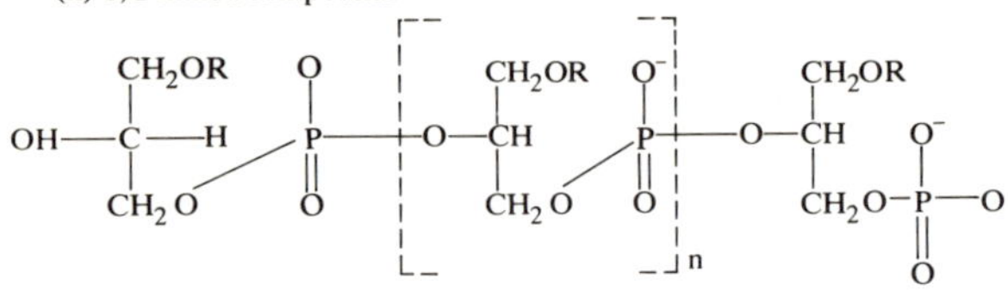

R in *Bacillus subtilis* var niger WM is β-glucosyl
in *B. stearothermophilus* is α-glucosyl
in *Actinomyces antibioticus* is O-α-D-galactopyranosyl-(1, 4)-α-D-*N*-acetylgalactosaminyl

(b) Ribitol teichoic acids

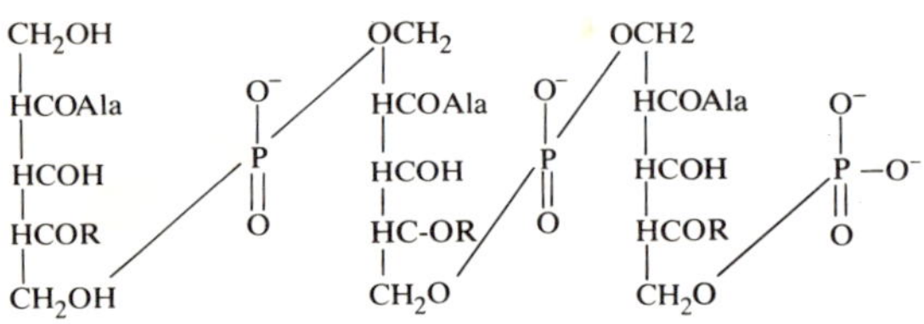

R in *Staphylococcus aureus* is either α or β-*N*-acetylglucosaminyl
in *Bacillus subtilis* strain W23 is α-glucosyl

(c) Compounds with intercalated glycosyl residues

(i) As in *Bacillus licheniformis* strain ATCC 9945

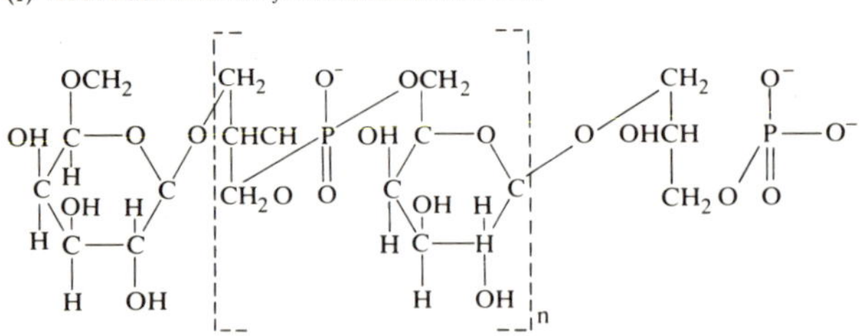

(ii) As in the capsular substances of *Diplococcus pneumoniae* type 13

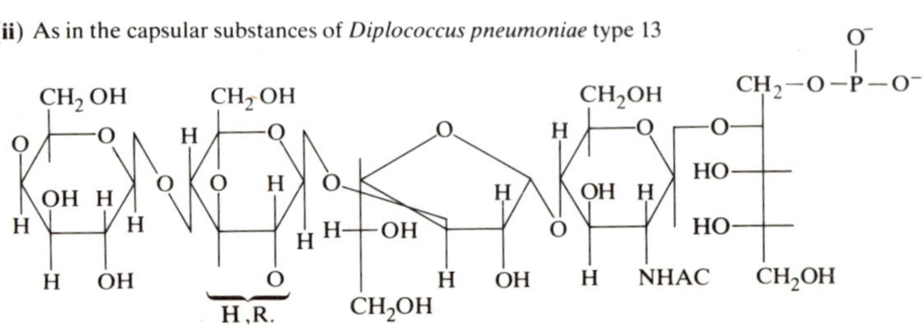

Fig. 2.7 Structures of teichoic acids.

peptidoglycans from different species vary in the extents to which the peptides are cross-linked. An index of cross-linkage is given by the ratio of the amino acid, R_3, having bound amino groups to the total R_3 residues in the peptidoglycan. The index can vary from 93.5% for *Staphylococcus aureus* to 20 to 30% for *Escherichia coli*. From this diversity Schleifer and Kandler have speculated about the taxonomic and evolutionary significance of peptidoglycan structure and a possible phylogenetic tree of bacterial evolution has been constructed. The length of the glycan chains in peptidoglycans has been subject to debate but it is now clear that some of the initial claims for rather short chain lengths were artefactual and caused by action of the autolytic enzymes in the organism. Present estimates suggest that in bacilli they are unlikely to be shorter than about 56 to 80 disaccharides long (i.e. 50 to 80 nm long), although in cocci such as staphylococci they may be quite short, about 9 to 10 disaccharides long. Little is known experimentally about the conformation of the polymers in walls apart from the presence of a considerable amount of hydrogen bonding deduced from infra-red spectroscopy. Diffuse X-ray patterns have been obtained but apart from the likelihood that the glycan strands run parallel with the surface of the bacteria examined, no further deductions can safely be drawn.

The teichoic acids The walls of nearly all Gram positive bacteria, unless grown under phosphate limitation, contain one or more of the polyol phosphate compounds collectively known as teichoic acids. Early examples obtained from the walls by rather prolonged extraction with 5% trichloroacetic acid at 0°C, were polyribitol or polyglycerol phosphate, substituted by hexoses via either the α or β linkage of the sugar, and by D-alanine via an ester bond (see Fig. 2.7 and Table 2.3). As

Table 2.3 The components of some wall teichoic acids; all contain phosphorus and some contain D-alanine.

Species of organism	Polyol	Sugars or other substituents
Bacillus subtilis strain W23	ribitol	glucose
Bacillus subtilis strain 168	(a)* glycerol	glucose
	(b) glycerol	glucose *N*-acetylgalactosamine
Bacillus licheniformis strain 6346	(a) glycerol	glucose
	(b) glycerol	glucose, galactose
Staphylococcus aureus	(a) ribitol	β-*N*-acetylglucosamine
	(b) ribitol	α-*N*-acetylglucosamine
Micrococcus Sp. 2102	—	*N*-acetylglucosamine
Staphylococcus epidermidis	(a) glycerol	α-glucose
	(b) glycerol	β-glucose
Streptococcus mutans	glycerol	galactose
Streptococcus pneumoniae	glycerol	choline
Lactobacillus plantarum	ribitol	glucose
Lactobacillus buchneri	glycerol	glucose

* (a) and (b) indicate the presence of more than one teichoic acid in the walls from the same culture or from different strains of the same species.

Table 2.4 Teichuronic acids of the walls of Gram positive bacteria

Organism	Components of polymer	Repeating unit	Mode of attachment to peptidoglycan
Bacillus licheniformis 6346	*N*-acetylgalactosamine D-glucuronic acid	→ 4)-D-glucuronosyl (1, 3)-*N*-acetyl-galactosaminyl-(1 →	Acid labile, probably phosphodiester
Bacillus subtilis var. *niger*	As above	Not known	Not known
Mycrococcus lysodeikticus	D-glucose *N*-acetyle-D-mannosaminuronic acid	→ 4)-*N*-acetyl-D-mannos-aminuronosyl-(1, 6)-D-glucosyl-(1 →	di-*N*-acetylmannos-aminuronosyl-*N*-acetylglucosaminyl-1-phosphoryl-
Staphylococcus aureus T	*N*-acetyl-D-fucosamine *N*-acetyl-D-mannos-aminuronic acid	Not known	Not known
Bacillus megaterium M46	D-glucose Glucuronic acid Rhamnose	Not known	Not phosphodiester Glycosidic?
Corynebacterium poinsettiae	Rhamnose Glucoronic acid Galactose Mannose Pyruvic acid	Not known	Not known
C. betae	Rhamnose Glucuronic acid Fucose Mannose	Not known	Not known

work has progressed, however, a rather wide variety of polymers containing sugars, polyols and phosphates has been identified. The structures of some of these are shown in Fig. 2.7. It is common for bacteria to have more than one type of teichoic acid in their walls. These can be extracted from wall preparations either by the original method using dilute acids, or by alkalies such as 0.1 **N** NaOH acting at 37°C. They are covalently attached to the 6-hydroxy positions of the muramyl residues in the glycan chains of the peptidoglycan. Attachment is frequently not a direct one, and small linkage units have been recognized for teichoic acids in walls of staphylococci and bacilli. Biosynthetic studies have confirmed the necessity for these linkage groups to be added before the teichoic acid molecules can be attached. In the examples so far examined the linkage unit consists of one mole of *N*-acetylglucosamine and three molecules of glycerol. The amino sugar is linked to the muramyl 6-OH by a phosphodiester bond, the glycerol molecules are linked by phosphodiester linkages, and finally the teichoic acid extends from this terminal glycerol.

The functions of wall teichoic acids in the economy of bacteria are unclear. As strongly negatively charged molecules they have a considerable affect on the surface charge of bacteria and they could have a role in regulating the supply of cations to the membrane. The presence of D-alanine with a free $-NH_2$ group has allowed the formulation of attractive hypotheses for the handling by the cells of divalent ions such as Mg^{2+}. Likewise the presence of these molecules in the wall has been shown to affect the expansion and contraction of walls that occur with alterations of pH and ionic strength in the environment. Conditional mutants with grossly reduced amounts of teichoic acids grow with aberrant morphologies. The teichoic acids also have a number of roles in the interaction of Gram positive bacteria with their environment. For example, absorption of some bacteriophages by *S. aureus* and strains of *B. subtilis* is dependent upon the presence of glycosylated teichoic acids attached to the peptidoglycan. In species like the lactobacilli the wall teichoic acids are the dominant cell antigens. Repeated attempts have failed to produce evidence for the presence of this class of polymer in Gram negative species of bacteria, although an exception to this general rule would appear to be the rumen micro-organism *Butyrivibrio fibrisolvens*. Walls of Gram positive bacteria can also contain a variety of other phosphate–containing polymers, such as those made of repeating units of 1–6 linked *N*-acetylglucosamine-1-phosphate and 3-0-α-glucopyranosyl-*N*-acetylgalactosamine-1-phosphate in different species of micrococci. These do not contain glycerol or ribitol but are usually nevertheless referred to as teichoic acids.

The teichuronic acids This group of acid polysaccharides was recognized first as a minor component of the walls of *Bacillus licheniformis* strain NCTC–6346 grown in batch culture, and subsequently in *Micrococcus luteus*. It has since been shown that when a number of species of bacteria are grown in the chemostat with inorganic phosphate as the limiting nutrient, teichoic acids cease to be formed and their place is taken by teichuronic acids. A number have been recognized (see Table 2.4). One component is always a uronic acid and may be either glucuronic acid, or the uronic acid derived from an amino sugar such as aminomannuronic or amino-glucuronic acids. Like the teichoic acids they are linked covalently to the glycan chains in the peptidoglycan by a single bond, but unlike them, linkage groups have not so far been demonstrated. The 6-hydroxy group of the muramyl residues in the glycan may not be universally involved; the teichuronic acid molecules in *M. luteus* walls, for example, are linked to the *N*-acetylglucosaminyl residues. When acid labile phosphodiester bonds are involved they are probably directly attached to the sugar

(a) In *Bacillus licheniformis* and *B. subtilis*

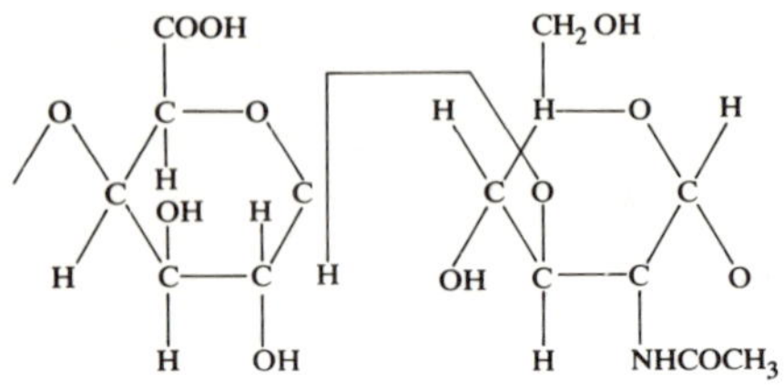

(i) 4–D–glucuronosyl-α-1 ⟶ 3 *N*-acetylgalactosaminyl 1 ——

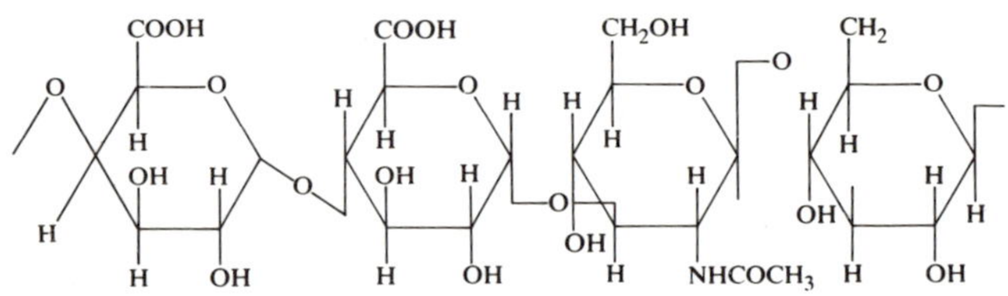

(ii) β ⟶ 4 GlcU $\xrightarrow{\beta 1,4}$ Glcu $\xrightarrow{\beta 1,3}$ Gal NAc $\xrightarrow{\alpha 1,6}$ Gal NAc ——

(b) In *Micrococcus luteus*

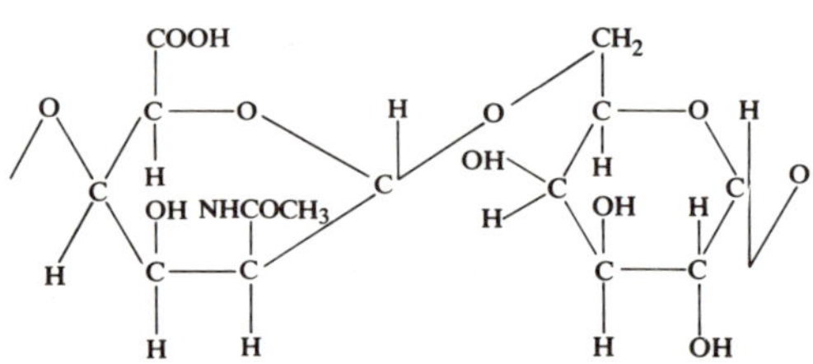

4-*N*-acetyl-D-mannosaminuronosyl-β-(1-6)-glucose

Fig. 2.8 Structures of teichuronic acids.

Table 2.5 Other wall polysaccharides yielding uronic acids on hydrolysis and likely to be teichuronic acids

Micro-organism	Recognized components
Staphylococcus aureus T	*N*-acetyl-D-fucosamine
	N-acetyl-D-mannosaminuronic acid
Bacillus megaterium M46	D-glucose
	Glucuronic acid
	Rhamnose
Corynebacterium poinsettiae	Rhamnose
	Glucuronic acid
	Galactose
	Mannose
	Pyruvic acid
Corynebacterium betae	Rhamnose
	Glucuronic acid
	Fucose
	Mannose

hydroxyl residues (e.g. in the walls of *B. licheniformis*, *B. subtilis* and *M. luteus*). In other species the relative acid-resistance of the bonds makes phosphodiester linkages less likely, and further work is necessary.

The structures of only two wall teichuronic acids are known (Fig. 2.8), and even one of these has recently been challenged, although polysaccharides that yield uronic acids among their hydrolysis products have been obtained from a number of organisms (see Table 2.5).

The lipopolysaccharides (LPS) This very large group of substances consist of a specific lipid – (lipid A) – buried in the outer membrane of the walls of Gram negative bacteria (Fig. 2.9). Complex polysaccharides containing unique sugars are attached to it (Figs. 2.10, 2.11). The importance of the lipopolysaccharides in the interaction of the bacteria with their hosts is clearly demonstrated by their role as O-somatic antigens in pathogenic salmonellae. The O-somatic antigens present in all Gram negative species are basic cellular antigens that are not changed by cultural conditions. They are of immense value in the classification of organisms like salmonellae. Their potent antigenicity has allowed them to be visualized under the electron microscope by first treating bacteria with antibodies labelled with ferritin before sections are cut. By this means it can be seen that they often protrude long distances from the outer membrane of the envelope.

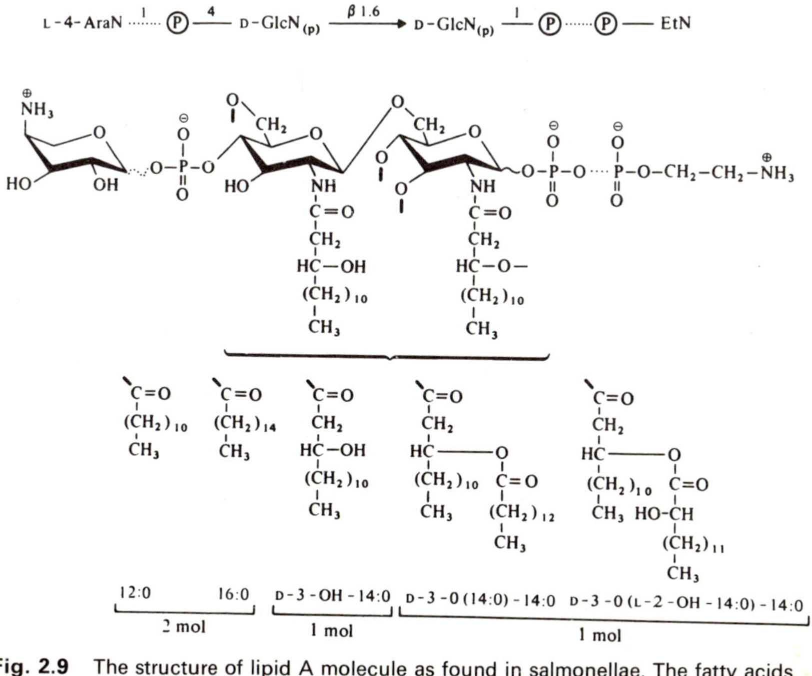

Fig. 2.9 The structure of lipid A molecule as found in salmonellae. The fatty acids and hydroxyacids in the lower part of the figure are found esterified to the amino groups of the glucosamine molecules. At the top of the figure is a simplification of the structure of lipid A. (from Rogers *et al.*, 1980)

Table 2.6 Structure of the 0-specific repeating units of *Salmonella* lipopolysaccharides (from Rogers *et al.*, 1980)

Salmonella group (figures in parentheses represent the O-antigenic specificities)		Repeating unit	
Group A	*S. paratyphi* A (1, 2, 12)	α-Par α-Glc |1, 3 |1, 4 →2-α-Man-1, 4-α-Rha-1, 3-α-Gal-1→ | Ac	*
Group B	*S. typhimurium* (4, 5, 12)	Ac-0-2-α-Abe α-Glc |1, 3 |1, 4 →2-α-Man-1, 4-β-Rha-1, 3-α-Gal-1→	*
	S. bredeney (wild) (1, 4, 12)	α-Abe α-Glc |1, 3 |1, 6 →2-α-Man-1, 4-β-Rha-1, 3-α-Gal-1→	*
Group C_1	*S. thompson* $(6_2, 7)$	Glc |1, 3 →Man-1, 2-Man-1, 2-Man-42-Man-1, 3-Gal→	†
Group C_2	*S. newport* $(6_1, 8)$	α-Abe α-GlcAc |1, 3 |1, 3 →4-α-Rha-1, 2-α-Man-1, 2-α-Man-1, 3-α-Gal→ |2 Ac	†
Group C_3	*S. kentucky* (8, 20)	α-Abe GlcAc |1, 3 |1, 4 →4-α-Rha-1, 2-Man-1, 2-Man-1, 3-Gal→	†
Group D_1	*S. typhi* (9, 12)	α-Tyv Ac-0-2-α-Glc |1, 3 | →2-α-Man-1, 4-α-Rha-1, 3-α-Gal-1→	*
Group D_2	*S. strasbourg* (9, 46)	α-Tyv α-Glc |1, 3 | →6-β-Man-1, 4-α-Rha-1, 3-α-Gal-1→	*

Table 2.6 (*continued*)

Salmonella group (figures in parentheses represent the O-antigenic specificities)		Repeating unit	
Group E_1	*S. muenster* (3, 10)	Glc |1, 4 →6-β-Man-1, 4-α-Rha-1, 3-α-Gal→	*
Group E_2	*S. newington* (3, 15)	α-Glc |1, 4 →6-β-Man-1, 4-α-Rha-1, 3-β-Gal→	*
Group E_4	*S. senftenberg* (1, 3, 19)	α-Glc |1, 6 →6-β-Man-1, 4-Rha-1, 3-α-Gal-1→	*
Group G	*S. friedenau* (13, 22)	Glc-1→? | →β-Gal-1, 3-Gal*N*Ac-1, 3-Gal*N*Ac-1, 4-Fuc→	†
Group L	*S. minnesota* (21)	α-Gal α-GlcNAc | | →β-Gal-1, 3-Gal*N*Ac-1, 3-Gal*N*Ac→	†
Group N	*S. godesberg* (30)	Glc |1, 4 →β-Glc-1, 3-Gal*N*Ac-1, 4-Fuc→	†
Group U	*S. milwaukee* (4, 3)	α-Gal |1, 3 →β-Gal-1, 3-Gal*N*Ac-1, 3-Glc*N*Ac-1, 4-Fuc→	†

*biological unit † chemical unit

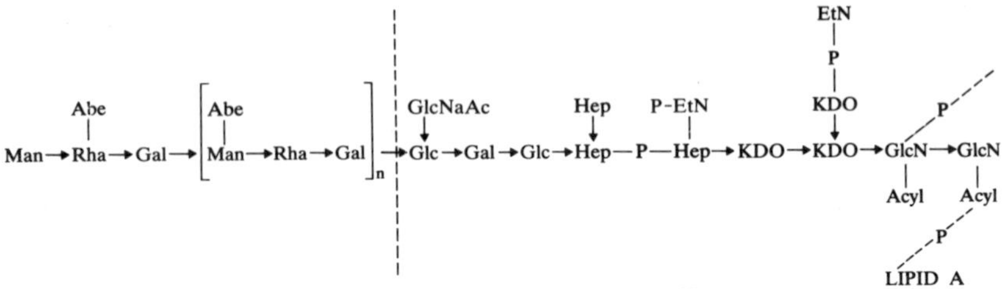

Fig. 2.10 The structure of lipopolysaccharide from *Salmonella typhimurium*
Abe = Abequose, Hep = L-glycero-D-mannoheptose (see Fig. 2.11), KDO = 2-keto-3-deoxyoctonic acid (see Fig. 2.11), EtN = ethanolamine. The remaining symbols have their usual meaning. For structure of lipid A see Fig. 2.9.

The chemical structures of the LPS of the salmonellae have been subjected to particularly intense investigation in relation to their immunological activities. The result is that large numbers of chemotypes have been recognized matching the equally large variety of immunological types. The chemical differences specifying this diversity are to be found in the polysaccharide side chains of the LPS (see Table 2.6). A whole molecule of LPS is built up of a core region and a side chain (see Fig. 2.10). The core region is attached to the lipid A (see Fig. 2.9). Although the most comprehensive work has been done on the chemical structures of the LPS from the salmonellae, partial and in some cases complete structures are also available from materials isolated from a wide variety of other Gram negative bacteria,

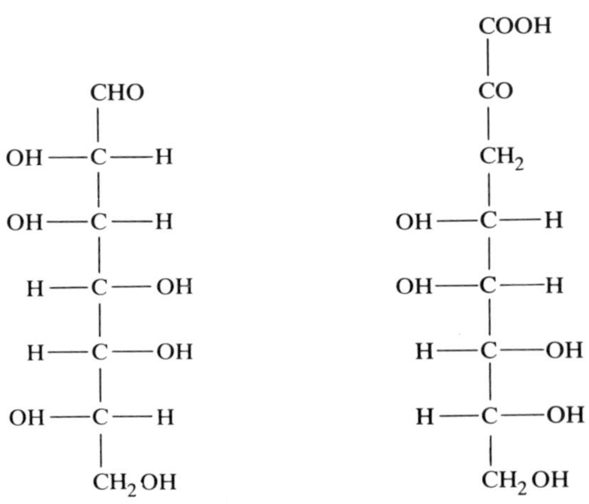

Fig. 2.11 The structure of two unique sugars of lipopolysaccharide.

among which may be mentioned Escherichiae, Shigellae, Klebsiellae, *Proteus* and Pseudomonadaceae, and there are strong indications for their presence on the surfaces of all Gram negative species. The general structures and components appear to have much in common in all species although it is of interest that both heptose and KDO are missing from LPS of the Bacteroides.

Waxes and lipids The amounts of lipid-soluble components in the walls of most Gram positive species are so small as to be regarded as contaminants from remaining membrane fragments. In Gram negative bacteria, of course, lipids are present in the outer membrane. The mycobacteria, nocardia and corynebacteria differ, however, in having a Gram positive type of ultrastructure but walls with 25 to 30% of their weight as lipid extractable by organic solvents. The walls also contain polysaccharide which is an arabogalactan esterified with complex long-chain fatty acids– the mycolic acids (see Table 2.7). The arabogalactan is covalently linked *via* a phosphate group to the 6-OH position of the *N*-glycollylmuramic acid residues in the glycan strands of the peptidoglycan. In some species of mycobacteria such as *M. tuberculosis* a fraction containing a tetrasaccharide of peptidoglycan joined to

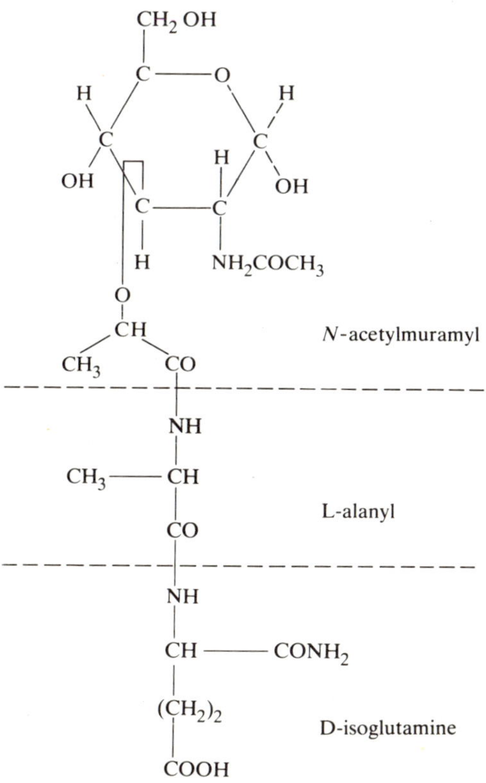

Fig. 2.12 The so-called minimum adjuvant structure (MDP).

Table 2.7 Mycobacterial mycolic acids

Strain	Principal mycolic acid	Structure
M. tuberculosis var. *hominis*	α-Mycolic acid	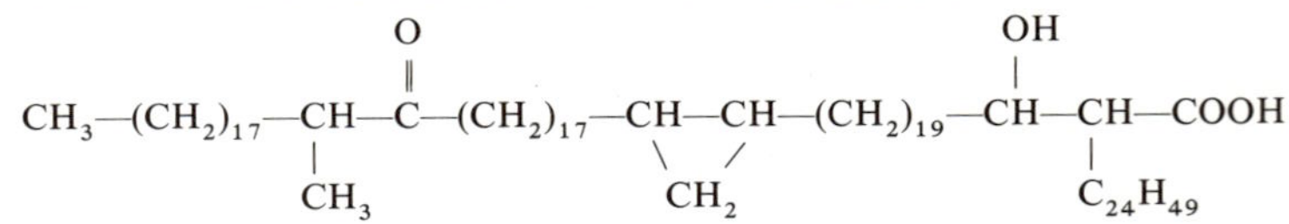
	Methoxylated mycolic acid	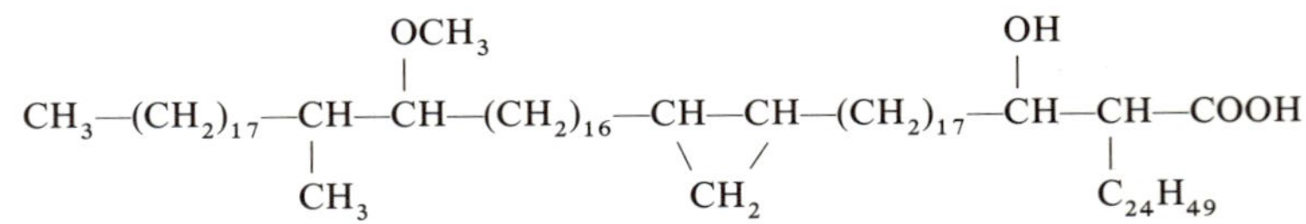
M. kansasii	α-Kansa mycolic acid	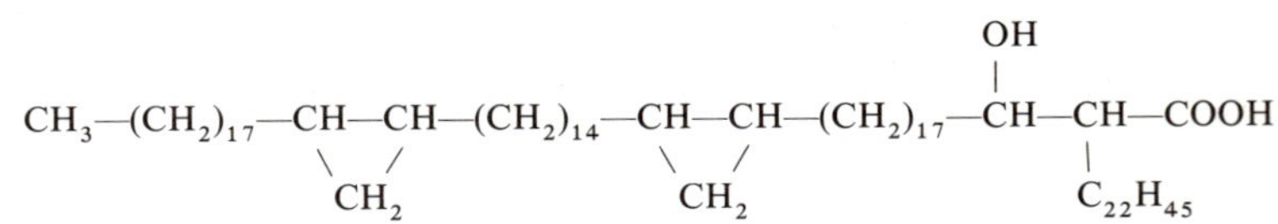
M. smegmatis	α-Smegmamycolic acid	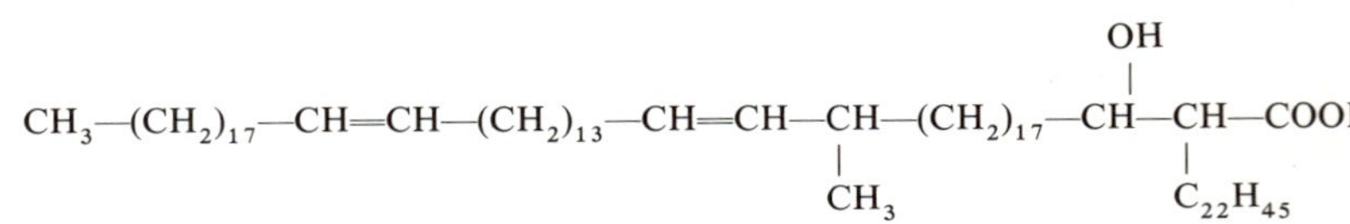
M. phlei	Dicarboxylic mycolic acid	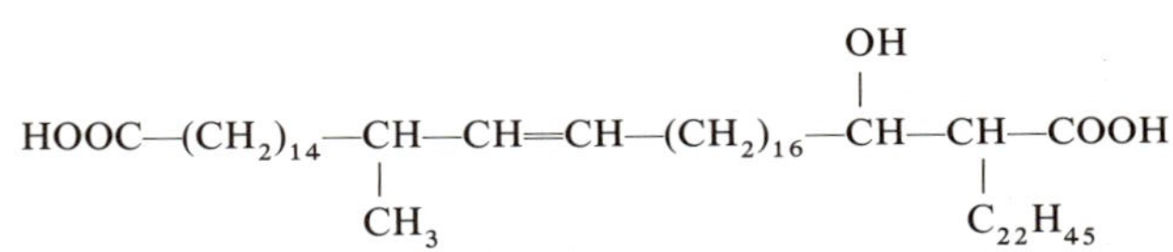

esterified arabogalactan can be isolated in a soluble form by extraction of whole bacteria. It is a major component of the so-called wax D fraction and probably represents an autolysis product rather than a truly soluble peptidoglycan. Similar fractions can be isolated from the walls of other mycobacteria by the action of lytic enzymes. Such fragments have great importance as immunoadjuvants. During the course of studies of wax D it has been shown that a simple peptidoglycan fraction (Fig. 2.12) that can be isolated from many organisms also has strong immunoadjuvant activity.

The archaebacteria

One of the most exciting recent developments has been the discovery of a new class of supporting polymers in the walls of *Methanobacterium*, which has the orthodox envelope ultrastructure of a Gram positive bacterium. These polymers have the general overall structure of the peptidoglycans in that glycan strands are cross-linked by short peptides which include isoglutamic acid, but they contain no muramic acid. The glycan strands, for example, in *Methanobacterium thermoantotrophicum* contain residues of *N*-acetylglucosamine and *N*-acetyltalosaminuronic acid. The carboxyl groups of the latter are linked in a manner analogous with the 3-lactyl carboxyls of *N*-acetylmuramyl in orthodox peptidoglycans and have short cross-linked peptides containing isoglutamyl, alanyl and lysyl residues (Fig. 2.13). Also present in these walls are polysaccharides yielding a variety of hexoses on hydrolysis. Other genera of the methanogenic bacteria such as the methanococci yield no insoluble preparation when orthodox means, such as treatments with sodium dodecyl sulphate and proteolytic enzymes, are used to isolate envelopes. It has been concluded that the walls are probably mostly of protein.

The novel properties of the walls of these organisms together with other taxonomic criteria such as sequence analysis of 16S rRNA have prompted the suggestion that three urkingdoms should be recognized: (1) the eubacteria with peptidoglycan-

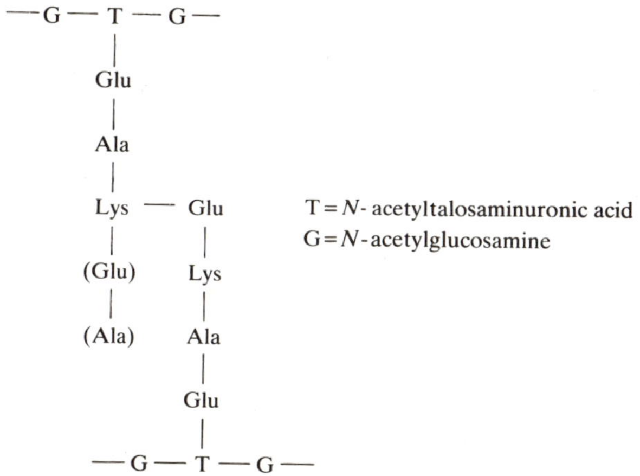

Fig. 2.13 The pseudomurein in cell walls of *Methanobacterium thermoautotrophium*.

containing walls, (2) 'that aspect of the eukaryotic cells represented in the 16S rRNA' and (3) the archaebacteria which would include the methanogens, the halophiles and the two thermo-acidophyles, *Sulpholobus* and *Thermoplasma*.

In those representatives of the methanogenic bacteria that have walls which are soluble in detergents such as sodium dodecyl sulphate, hydrophobic interactions between the component proteins or glycoproteins may play an important part in maintaining their integrity. Envelopes of halobacteria, on the other hand, which also contain no peptidoglycan, disintegrate if the very high salt concentrations required by the bacteria for growth (of the order of 5 to 10 M) are diluted. This suggests that ionic interactions are of major importance for these walls. Ultrastructurally regular patterns can be seen on the surface of the wall layer which can be distinguished from the underlying membrane. When the ionic strength of the surrounding environment of the bacteria is lowered spheres are formed from rod-shaped micro-organisms and these eventually lyse. This is not simply an osmotic effect, however, since wall preparations also disintegrate under these conditions. A high molecular weight glycoprotein has been isolated from the walls of *Halobacterium salinarium* which makes up 50% of their weight and accounts for all the non-lipid carbohydrate. The carbohydrate attached to each molecule of the protein consists of one large heterosaccharide. This contains glucose, galactose, mannose, glucosamine and an unidentified amino sugar, linked by a glycosylamine linkage to asparagine, together with about 20 disaccharides and 12 to 14 trisaccharides. Both types of oligosaccharides are attached by alkali-labile O-glycoside linkages. These types of linkages are characteristically present in glycoproteins from eukaryotic rather than prokaryotic sources. The presence of a very high proportion of glutamyl and aspartyl residues in the protein and of uronic acid groups in the carbohydrate account for the strongly acidic charge on the glycoprotein, and hence possibly for the necessity of a high ionic concentration to maintain the integrity of the walls by shielding the negative charges and reducing mutual repulsion. Very interestingly bacitracin prevents growth of *H. salinarium*, but first the rods become spheres even when growing in 25% NaCl. Bacitracin inhibits dephosphorylation of isoprenoid lipid intermediates involved in the biosynthesis of many bacterial sugar-containing polymers such as lipopolysaccharides, peptidoglycan and the mannan of *Micrococcus luteus*. The glycosylation of the glycoprotein in the envelopes of *H. salinarium* is also diminished. It is, therefore, reasonable to suppose that the carbohydrate chains attached to the protein play a role in maintaining the shape of the rods.

Summary

The morphology of the outer layers of Gram positive and Gram negative walls differs as does their chemistry. In the former the walls are relatively thick structures consisting of peptidoglycans, teichoic and teichuronic acids and polysaccharides. In the latter the wall has several layers which include a very thin layer of peptidoglycan and an outer membrane with lipopolysaccharides attached to it. The peptidoglycans of both Gram positive and Gram negative bacteria consist of glycan strands joined together by short peptides made up of alternating L and D amino acids. Only a few types of amino acid are present in the peptidoglycan of any one species of micro-organism, but over a range of bacterial species a large number is present, some unique to the polymers; D-alanine is always present. Four or five different types of peptidoglycan are known, differing in their cross-linking. Peptidoglycans from different Gram negative bacteria have identical structures and have lipoprotein molecules covalently attached by single bonds. In Gram positive species there are a range of phosphate-containing polymers – the teichoic acids, acidic polysaccharides – the teichuronic acids, and polysaccharides. The peptidoglycan in Gram negative species has covalently attached to it a unique hydrophobic polypeptide which stretches between this layer and the outer membrane. Lipopolysaccharides are fixed to the outer leaflet of the outer membranes by the lipid A. A wide variety of chemotypes each with a unique lipopolysaccharide are to be found among extensively studied micro-organisms such as the salmonellae where each is related to the serotype. Some of the archaebacteria have a pseudo-peptidoglycan which, although constructed on the same general plan as common peptidoglycans, has different components.

References

General

ROGERS, H. J., PERKINS, H. R. and WARD, J. B. (1980). In: *Microbial Cell Walls and Membranes*. Chapman and Hall, London and New York. Chapters entitled: Ultrastructure of bacterial envelopes, Isolation of walls and membranes, Structure of peptidoglycan, and Additional polymers in bacterial walls.

Peptidoglycans

GHUYSEN, J-M (1968). Use of bacteriolytic enzymes in the determination of wall structure and their role in cell metabolism. *Bacteriological Reviews* 32: 425–64.

ROGERS, H. J. (1974) Peptidoglycans (mucopeptides): structure function and variations. *Annals of the New York Academy of Sciences* 235: 29–51.

SCHLEIFER, K. H. and KANDLER, O. (1972). Peptidoglycan types of bacterial cell walls and their taxonomic implications. *Bacteriological Reviews* 36: 407–77.

Teichoic acids

BADDILEY, J. (1972). In: *Essays in Biochemistry* Vol 8, pp. 35–78. Edited by P. N. Campbell and F. Dickens. Academic Press, London and New York.

Lipopolysaccharides

LUDERITZ, O., WESTPHAL, O., STAUB, A. M. and NIKAIDO, H. (1971). In: *Microbial Endotoxins* Vol 4, pp. 145–223. Edited by G. Weinbaum, S. Kadis and S. J. Ayl. Academic Press, London and New York.

WILKINSON, S. G. (1977). Composition and structure of bacterial lipopolysaccharides. In: *Surface Carbohydrates of the Prokaryotic Cell* pp. 97–175. Edited by I.W. Sutherland. Academic Press, London, New York and San Francisco.

Mycobacteria

BARKSDALE, L. and KIM, K. S. (1977). Mycobacterium. *Bacteriological Reviews* 41: 217–372.

PETIT, J–F and LEDERER, E. (1978). Structure and immunostimulant properties of mycobacterial cell walls. *Symposium of the Society for General Microbiology* 28: 177–99.

Patterned layers

SLEYTR, U. B. (1978). Regular arrays of macromolecules on bacterial cell walls: structure, chemistry, assembly and function. *International Review of Cytology* 53: 1–64.

3 The membranes of bacteria

The internal organization of the vast majority of bacteria is undoubtedly simpler than that of unicellular eukaryotes such as *Chlamydomonas*, yeasts or amoebae. Specialization in morphological terms has often been reduced to the minimum. A word of caution, nevertheless, may be necessary, since visible specialization is very much the product of the technology available. Whilst visibly differentiated mitochondria, Golgi apparatus, or nuclear membranes may be missing in bacteria, topological specialization of the few membranes present may still occur but be beyond the limits of our present means of observation. Considerable differentiation of the outer membrane of Gram negative species has already been revealed by biochemical and physiological probing but this is not apparent morphologically.

In a very large number of species of bacteria, including those used as workhorses in the laboratory such as *Escherichia coli*, *Bacillus subtilis* and *Streptococcus faecalis*, the only membranes underlying the wall are the cytoplasmic and mesosomal ones (see Fig. 3.5). Moreover the latter can only be seen in stained and fixed preparations as we shall see (p. 38) and are therefore in a sense artefacts. Photosynthetic, very active nitrogen-fixing, and methane-utilizing organisms as well as the eyanobacteria have large amounts of internal membranes but often in rather simple configurations (see Fig. 3.1).

The currently accepted model for the structure of membranes in both prokaryotes and eukaryotes is the fluid mosaic model proposed by Singer and Nicholson (see Fig. 3.2). Briefly, the phospholipids are arranged as a bilayer with their hydrophobic fatty-acid chains facing inwards and their hydrophilic heads facing outwards. Into this lipid sea the proteins are inserted to various depths. Two classes of proteins can be recognized. The intrinsic ones are held by hydrophobic bonds, and can only be rendered soluble by dissociating the whole membrane structure, usually with neutral detergents such as deoxycholate. The other class of proteins is peripheral and can be removed by chelating reagents such as dilute solutions of ethylene diaminetetraacetic acid or even by washing with solutions of low ionic strength. These are presumably held to the membrane by ionic bonds and involve divalent cations which in bacteria are probably Mg^{2+}. A further sub-class of integral proteins are the transmembrane proteins which protrude on both sides of the membrane.

One of the important consequences of the fluid mosaic model is that it allows a relatively easy explanation of the known lateral movement of both phospholipids, which is rapid, and proteins, which is slow. Such movement may be of great importance in topological disposition within the membrane of enzymes that biosynthesize wall polymers, particularly during growth and division of bacteria.

The cytoplasmic membrane

Although bacterial membranes have compositions and properties that will distinguish them from the membranes of mammalian cells, there is little to distinguish them morphologically. After fixing and staining bacteria by the usual methods, such as those employing osmium tetraoxide with or without prefixation with glutar-

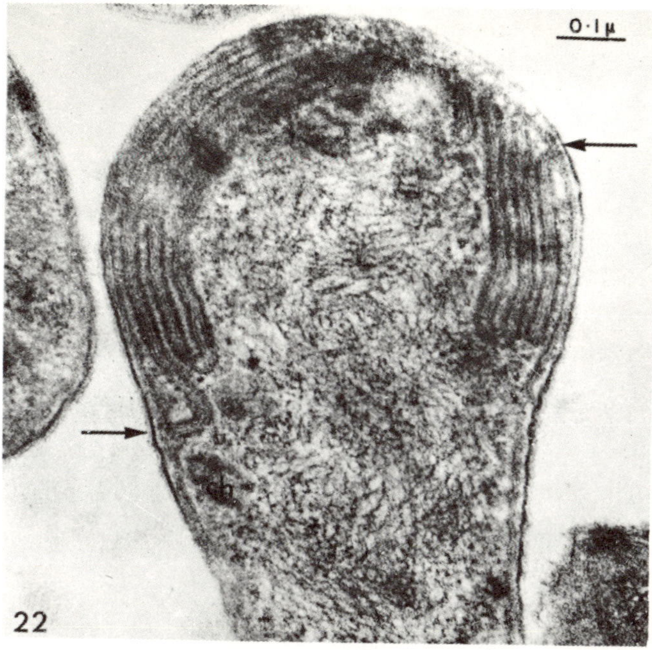

Fig. 3.1 A transverse section of *Nitrobacter agilis* showing layers of infolded cytoplasmic membrane. The ← shows the points at which the membranes can be seen to join the cytoplasmic membrane. (Photo: C.W. Watson)

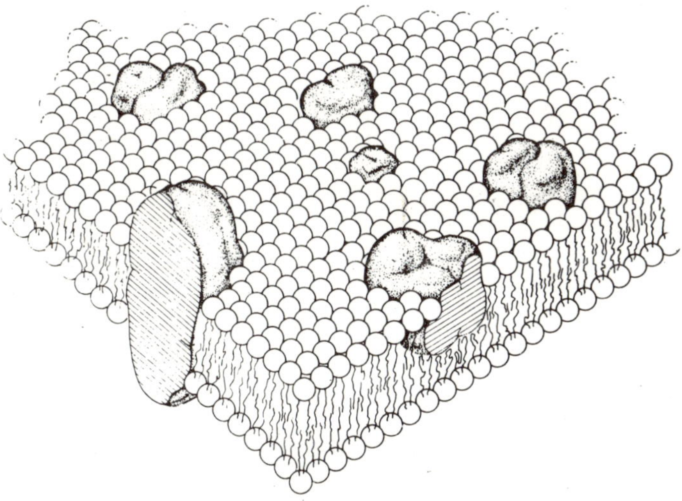

Fig. 3.2 The structure of biological membranes according to the fluid mosaic model. The empty circles are the hydrophilic heads of phospholipid molecules whilst the shaded shapes represent proteins set to various depths in the lipid layer.

Table 3.1 Subunit size and composition of the F_1-ATPases

Micro-organism	Molecular weights ($\times 10^{-3}$)						Subunit composition
	Enzyme	α	β	γ	δ	ε	
Rat liver mitochondria	340–380	53–62.5	50–57	25–36	12–12.5	7.5	$\alpha_3\beta_3\gamma\delta\varepsilon$
Yeast mitochondria	340	58	54	38	38	12	—
Escherichia coli	340–400	56–60	52–56	32–35	21	11–13	$\alpha_3\beta_3\gamma\delta\varepsilon$*
Salmonella typhimurium		57	52	31	21	13	$\alpha_3\beta_3\gamma\delta\varepsilon$
Alcaligenes faecalis	350	59	54	43		12	—
Thermophilic bacterium	380	56	53	32	11	15	—
Streptococcus faecalis	345	60	55	37	20	12	—
Bacillus megaterium	399	68	65	—	—	—	$\alpha_3\beta_3$
Micrococcus luteus	345	52–60	47–60	41	28		$\alpha_3\beta_3\gamma(\delta\varepsilon)$†
Bacillus subtilis	315	59	57	—	—	—	$\alpha_3\beta_3$

* One claim for a composition of $\alpha_2\,\beta_2\,\gamma_2\,\delta_2\varepsilon_2$

† Presence of δ and ε not yet settled

aldehyde, sections examined by the transmission electron microscope show a so-called *double track* membrane about 5 to 7 nm in thickness consisting of two dark bands sandwiching a less opaque region. When bacteria are harvested quickly from rapidly growing cultures and carefully fixed, stained and dehydrated, this membrane, which packages the cytoplasm is seen to be intimately adjacent to the wall in Gram positive organisms or to the thin rigid layer in Gram negative species (see Fig. 3 Chapter 2). Indeed, it seems that the structures are spot welded together in Gram negative organisms at many points per cell. In Gram positive species plasmolysis is more difficult to achieve, which may mean that the membranes are fixed to the walls at more numerous points. Various attachment structures have been claimed to to be seen, ranging from arrays of peg-like structures to strands of membrane and 'globs' of material. When bacteria are freeze-fractured (see p. 8) one of the fracture planes is within the fatty interior of the membrane bilayer. This splits the membranes into two halves, one having a concave and the other a convex shape. As with all membranes the replicas of one face show a random arrangement of approximately spherical bodies. It is generally agreed that these are proteins. The complementary face shows a random array of pits each corresponding to a body on the other face. The only remarkable feature of bacterial membranes as compared with those from other sources is that the particles are more densely packed. This would be expected from their higher protein content.

The cytoplasmic membranes can readily be isolated from Gram positive organisms by suspending them in strong solutions of sucrose and then removing the walls with any suitable enzyme that hydrolyzes peptidoglycan. If the walls are removed from bacteria whilst they are suspended in a normal culture medium or in any other solutions of low osmolarity the unsupported cell membranes immediately rupture and the cells lyse. This is because there is a high internal concentration of low molecular weight metabolites in bacteria; water rushes through the cytoplasmic membrane, as it would through a cellophane dialysis sac, and if there is no wall to prevent the cell swelling it will continue to do so until the cytoplasmic membrane is broken. If the external medium contains a high concentration of substances, like sucrose, that cannot rapidly penetrate the membrane, this does not happen and the cell does not burst. The resulting membrane-bound bodies suspended in sucrose are called protoplasts. Flagella usually remain in suspension although they sometimes are fixed to the protoplasts. The mesosomal membrane (see p. 35) is extruded. These various entities can be separated by differential and gradient centrifugation. The larger protoplasts can first be deposited by gentle centrifugation, washed and lysed by dilution of the sucrose. The membranes are subsequently deposited by centrifugation at a higher speed, if necessary, after blending to remove flagella, and after the viscosity of the cytoplasmic DNA has been reduced by DNAase treatment. The whole process is carried out in the presence of Mg^{2+} to maintain the integrity of the membranes, although the concentration must be regulated carefully to prevent adhesion of the mesosomes to the cytoplasmic membrane. When such preparations of cytoplasmic membrane are negatively stained, the inner surface can be seen to be densely covered by white spots representing areas not penetrated by the stain. By manipulating the ionic environment, or by treating the membranes with dilute solutions of chelating agents such as EDTA, the spots can be removed, showing their peripheral nature. Biochemical analysis has shown them to be the organized subunits of the F_1 part of Ca-Mg activated ATPase. When removed from the membranes the subunits, usually five in number (see Table 3.1), can be re-assembled to form material of the same form as the original 'spots' or can be re-assembled on to the membranes from which they have been removed. The ATPase function of such membranes is fully restored, showing that the re-assembly is not a topologically random process but is determined by the positions of another part of the ATPase enzyme, the F_0 subunit, which remains behind with the membrane. The F_0 part of ATPase consists of integral proteins and appears to form a pore in the membrane through which protons can pass. The F_0 part is vitally concerned, in concert with F_1, in oxidative phosphorylation (see Fig. 3.3.).

The isolation of the cytoplasmic membrane from Gram negative species is necessarily more complicated than from Gram positive bacteria because of the presence of the separate outer membrane. Fortunately, there is a considerable difference in density between the cytoplasmic and outer membranes allowing them to be separated by sucrose gradient centrifugation. The usual procedure is to hydrolyze the peptidoglycan with lysozyme, a chelating agent also being present, and then, after mechanically disrupting the sphaeroplasts, to apply them to the top of a sucrose gradient and centrifuge to equilibrium. The density of the cytoplasmic membrane is 1.14 to 1.16 and that of the outer membrane is 1.22. An intermediate peak always appears with a density of 1.19; this is a mixture of inner and outer membranes. Fortunately the unique presence of components of the electron transport chain such as the dehydrogenases and cytochromes in the cytoplasmic membrane, and of lipopolysaccharide components such as KDO (see Chapter 2) or heptose in the outer membrane, allows a rather exact monitoring of the success of separation methods.

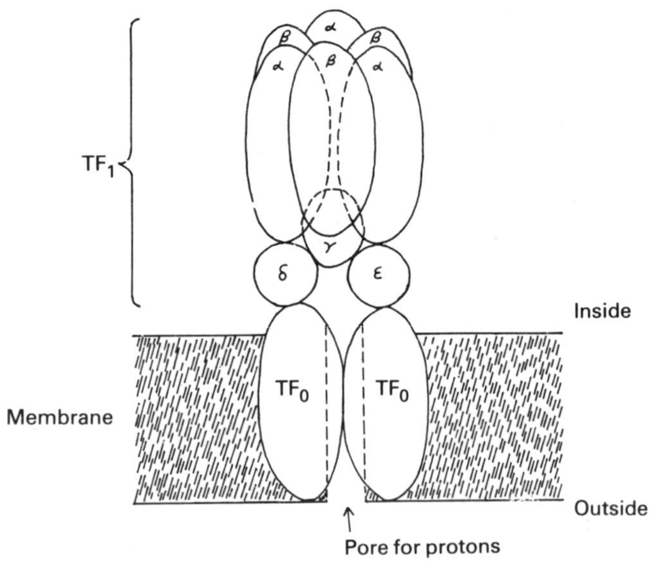

Fig. 3.3 The structure of ATP'ase when functionally attached to the membrane. The units α and β can alone hydrolyse ATP, whilst γ, δ and ε are involved in attaching the α and β units to the membrane surface. Together these form the peripheral F_1 part of the enzyme. This complex is then attached to integral membrane proteins constituting the F_0 part of the enzyme, which forms a pore through which protons can pass during oxidative phosphorylation.

The sidedness of the cytoplasmic membranes If the behaviour of whole cells or protoplasts towards a variety of substrates such as nucleotides that cannot penetrate membranes is compared with that of membranes disintegrated by, say, treatment with ultra-sound, fewer compounds are found to be attacked by the cells or protoplasts than by the broken preparations. The reason for this difference is that some enzymes like ATPase are confined to the inner side of the protoplast membrane. One of the neater and more general ways of examining the distribution of proteins across the cytoplasmic membranes of micro-organisms, Gram positive ones in particular, is by the use of immunological methods and crossed immunoelectrophoresis, or so called rocket immunoelectrophoresis, named from the shapes of the tracks obtained. In this technique the antigen mixture is first run in one direction in a narrow strip of gel, which is then placed against a slab of gel impregnated with antibody. Rocket-like tracks of immunoprecipitates formed by antigens reacting with antibodies are then found in the second gel at positions determined by the distance of migration of the antigens in the first gel; their concentrations determine the height of the rockets (see Fig. 3.4). If rabbits are immunized with whole membrane preparations from *E.coli* and the resulting antisera used, some 46 antigens can be resolved; since the whole process of crossed immunoelectrophoresis takes place in relatively harmless buffer and uses neutral detergents such as Triton X-100, enzymic activities can also be detected by incorporation of suitable substrates into the gels. The products of more than a dozen activities have been so identified. If instead of using the unmodified antiserum in the second gel, it is first absorbed with unruptured protoplasts, only those proteins which are exposed on the outside of the membrane combine with their antibodies and are removed. When such adsorbed serum is used for crossed immunoelectrophoresis, only those antigens in the membrane that did not react during the adsorption procedure will now show tracks in the gel. It is a reasonable assumption that these are proteins that face the cytoplasm in the original protoplast membrane. Trans-membrane proteins will presumably form exceptions to this

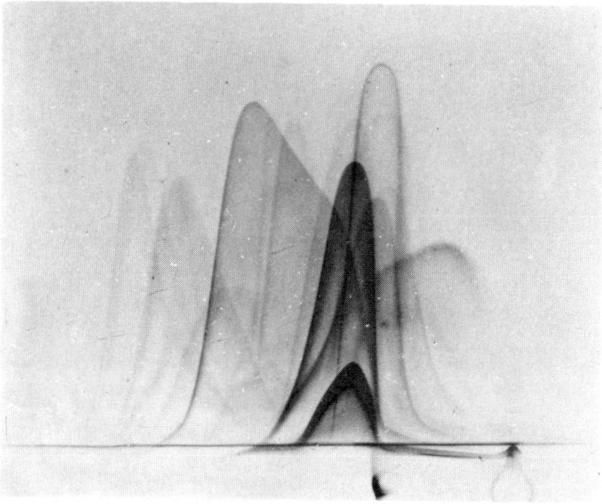

Fig. 3.4 Rocket immunoelectrophoresis of the cytoplasmic membranes from *Micrococcus luteus*. Each track represents a separate antigen (almost all proteins) present in the membranes. (Kindly supplied by Dr M.R.J. Salton)

Table 3.2 The orientation of bacterial membranes in vesicles

Micro-organism	Method of preparation	Method of examination	Orientation deduced
E.coli	Lysozyme-EDTA	Activities of succinic and glycerol 3-phosphate dehydrogenases	c. 50% inverted
		ATPase and reaction with antibody NADH-$K_3Fe(CN)_6$* reductase	c. 50% inverted
	Lysozyme-EDTA	D-lactate dehydrogenase reaction with antibody	85% uninverted
	Lysozyme-EDTA	Ultrastructure: Section and freeze-fracture	Uninverted
	Lysozyme-EDTA	Crossed immuno-electrophoresis	99% uninverted
	Lysozyme-EDTA	Ultrastructure: Freeze-etching	Uninverted
	Lysozyme-EDTA	Uptake of ^{3}H-vinylglycollate	Uninverted
	Lysozyme-EDTA, freezing and thawing	Ultrastructure: Freeze-etching	25% inverted
	EDTA-lysozyme	Antibody to ATPase	50% uninverted
	Lysozyme-EDTA and sonication	D-lactate dehydrogenase reaction with antibody	> 70% inverted
	Lysozyme-EDTA	ATPase, NADH dehydrogenase, ATPase antibody	Enzyme translocation
	French Pressure Cell	ATPase. Reaction with antibody $NADH_2$-$K_3Fe(CN)_6$ reductase	60–100% inverted
	Sonication	D-lactate dehydrogenase reaction with antibody. Ultrastructure: Freeze-etching ATPase antibody reaction, $NADH_2$-$K_3Fe(CN)_6$ reductase	All inverted
Mycobacterium phlei	Lysozyme	Cryptic oxidative phosphorylation	Uninverted
	Sonication	Expressed oxidative phosphorylation	Inverted
Bacillus subtilis	Lysozyme	Ultrastructure†: Freeze-etching	Uninverted
Micrococcus lysodeikticus	Lysozyme	ATPase reaction with antibody and ^{125}I-lactoperoxidase	At least partly inverted
	Lysozyme, Sonication	Accumulation of organic anions and K^+	Mixture of inverted and uninverted
Bacillus caldolyticus	Sonication	H-pump	Inverted

*$K_3Fe(CN)_6$ was used as an impenetrable anionic electron acceptor. It was not reduced in the presence of NADH by protoplasts or sphaeroplasts unless these were treated with either toluene or Triton X-100. When vesicles behaved similarly they were assumed to be surrounded by membrane of the same orientation as that on the protoplasts, i.e. uninverted.

† Vesicles within vesicles were seen and of these about 85% were uninverted.

rule. In membrane preparations from *Micrococcus luteus* 27 antigens were detected as present in membranes using non-adsorbed sera, whereas adsorption with protoplasts removed 12 which were presumably outward facing proteins. Other methods involving non-penetrating substrates, ^{125}I and ferritin-labelled antibodies, have also been used to distinguish inward and outward facing enzymes and proteins. As might be expected, many of the enzymes facing inwards are components of the electron transport chain such as cytochromes and the dehydrogenases. Some of the cytochromes of higher potential face outwards, reflecting the 'looped' arrangement of the respiratory system across the membrane. In general, less is known about the outward facing proteins.

The vesiculation of membranes One of the properties shown by all biological membranes is that of resealing into closed vesicles instead of remaining as open fragments accessible to all molecules on both surfaces. This property together with the anisotropic distribution of enzymes in membranes has led to much discussion, not all of it very fruitful, about the behaviour of membrane preparations both in studies of electron transport and of the active transport of molecular metabolites such as amino acids. One of the advantages of using membrane vesicles rather than whole bacteria or protoplasts for studying the transport of metabolites across membranes is the absence inside them of the machinery necessary for subsequent metabolism of the substances transported. It is, however, clearly important to know whether or not the membrane in the vesicles has the same orientation as that in the bacteria. The proof of this seemingly simple fact has been remarkably difficult to achieve. It has rather slowly emerged that if the membranes in protoplasts or whole bacteria are disintegrated by relatively violent means such as are involved in the use of ultrasound, or mechanical disintegration by one of the various machines designed for disrupting bacterial cells, the orientation of the membrane in the resulting vesicles is almost wholly the reverse of that present in the bacteria. If, however, sphaeroplasts, particularly of *E. coli* which has been thoroughly examined, are gently lysed, then a high proportion (some workers would say all) of the vesicles have membranes with the same orientation as those in the original bacteria. Table 3.2 summarizes some of the various studies and the methods that have been used in attempts to decide the orientation of membranes in vesicles. One feature that has not yet received a completely satisfactory explanation is that if more that one enzyme is used as an indicator of membrane orientation, they may not give the same answer. This has led to the hypothesis that some enzymes can migrate from their original situation during preparation of the vesicles and thus appear on the 'wrong' side of the membrane. However other explanations are possible and the problem is so far unresolved.

The mesosomes

When examined after the fixing and staining techniques discussed earlier, sections of bacteria such as *B. subtilis* or *S. faecalis* show characteristic organization of internal membranes seeming to be intrusions of the cytoplasmic membrane. These intrusions are the mesosomes (they have also been called chondroids and plasmalemmasomes). In bacilli, for example (see Fig. 3.5), which possess several mesosomes per cell, they appear to consist of globlets of intruding cytoplasmic membrane filled with tubules, small spheres (diameter 24 nm), or whorls of membrane. The appearance varies according to the fixing and staining procedures that have been applied. In

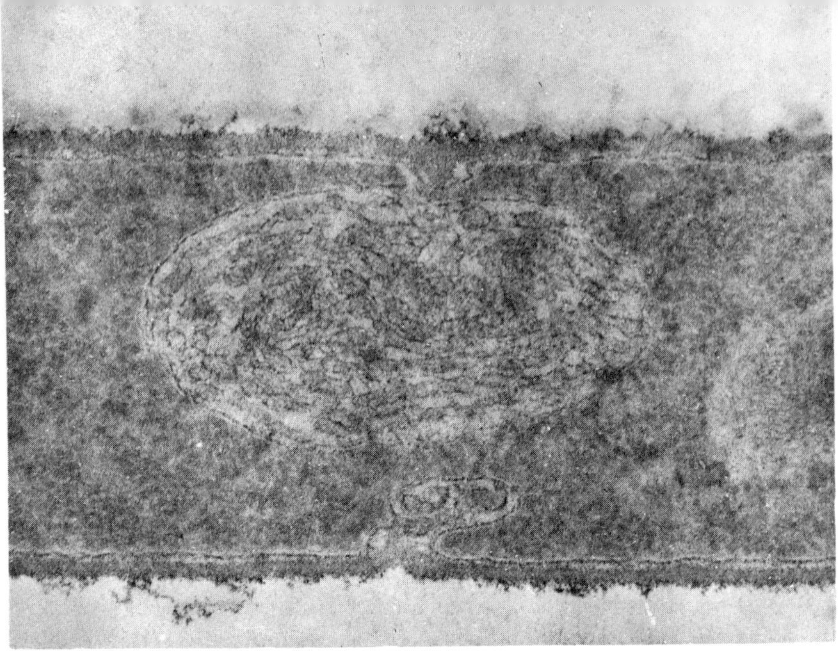

Fig. 3.5 The mesosome of *Bacillus licheniformis* strain 3646 as seen in a section of stained and fixed material. (Photo: Dr Ian Burdett)

Fig. 3.6 A negatively stained preparation of the cytoplasmic membrane of *B. licheniformis* with long branched strings of mesosomes still attached to it. (Photo: Dr Ian Burdett)

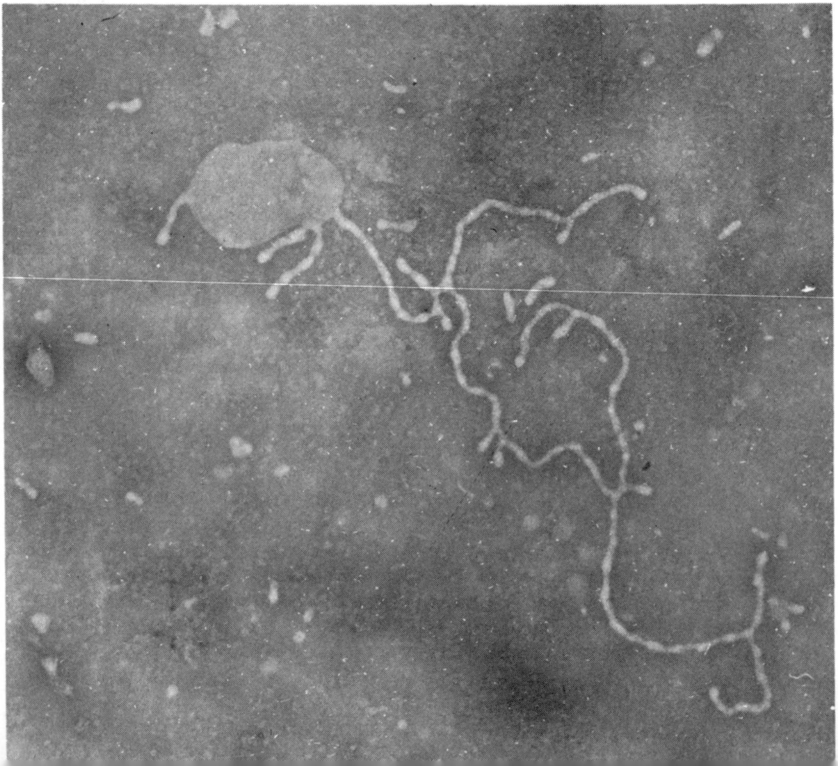

contrast, the single mesosome of streptococci appears to be made up of a long tube wound up to form a globlet-like structure. The largest mesosomal intrusions of the cytoplasmic membrane are universally associated with the site of division of bacteria although they sometimes also occur elsewhere. They are present in almost all species of bacteria, but are very much more prominent in Gram positive species – it has even been claimed that they are present in the mitochondria of eukaryotic micro-organisms. When Gram positive species are suspended in strong sucrose solutions the mesosomes are extruded into a depression of the cytoplasmic membrane under the wall. When the wall is removed to form protoplasts they stretch out in long tubules often having regular constrictions making them look like strings of pearls (Fig. 3.6). These can be readily separated from the protoplasts, providing the concentration of Mg^{2+} ions in the protoplasting medium is not too high; otherwise they stick firmly to each other and to the protoplasts. Low speed centrifugation removes the protoplasts and then high speed centrifugation deposits the mesosomes. If the organisms have flagella these too are present in the final deposit of mesosomes, along with membrane fragments arising either from a few ruptured protoplasts or from the surfaces of unruptured ones. It is, therefore, desirable to centrifuge the deposited mesosomal fraction through a gradient of either sucrose or caesium chloride. A band of 'purified' mesosomes can then be obtained. Such preparations consist either of vesicles (from *B. licheniformis*, Fig. 3.7) or tubules (*M. luteus*, Fig. 3.8).

Fig. 3.7 An electron microscopic section of the isolated mesosomal vesicles from *Bacillus licheniformis* (x 76,860). (Photo: Dr Ian Burdett) *left*

Fig. 3.8 A negatively-stained preparation of isolated mesosomal membranes from *Micrococcus luteus* (x 97,000). (Kindly provided by Dr John Freer) *right*

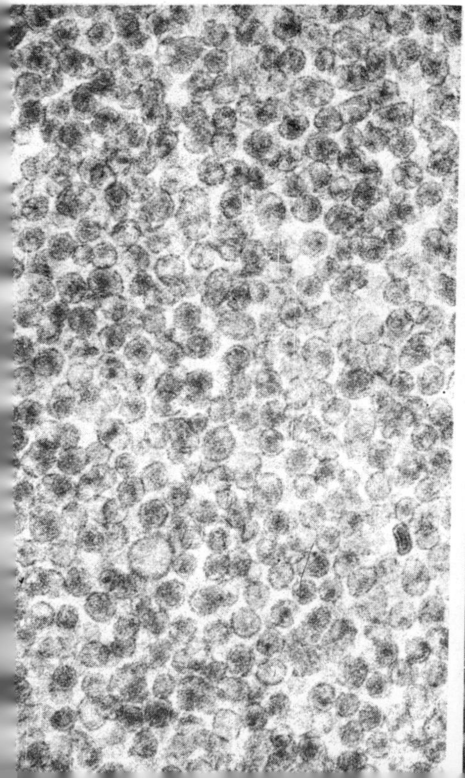

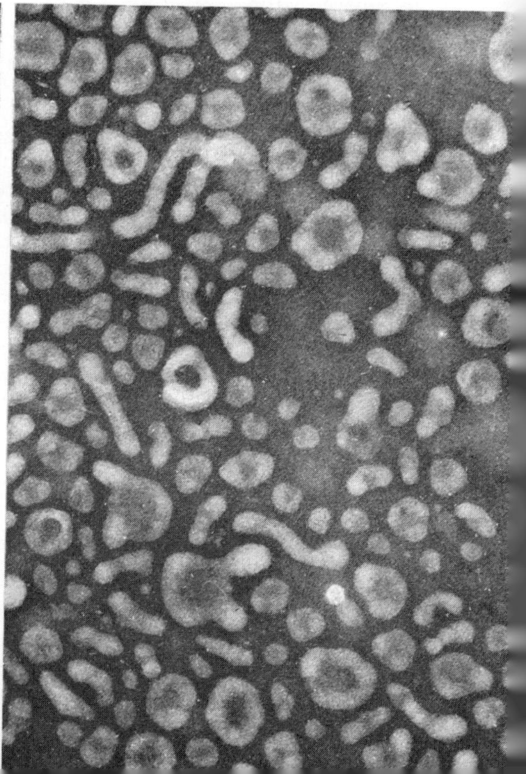

Table 3.3 Components of the electron transport systems absent or poorly represented in mesosomal membranes

Enzyme	Micro-organisms
Succinic dehydrogenase	*Bacillus subtilis, B. licheniformis, B. megaterium, Staphylococcus aureus, Micrococcus luteus*
NADH oxidase	As for succinic dehydrogenase also *Listeria monocytogenes*
Malate dehydrogenase	*B. megaterium, M. luteus, S. aureus*
α-glycerophosphate dehydrogenase	*S. aureus*
Cytochromes	As for succinic dehydrogenase
L-lactate dehydrogenase	*B. megaterium, M. luteus, S. aureus.*

Undoubtedly, in the above circumstances the mesosomal membrane is 'real' and has a number of differences from the cytoplasmic membrane. For example, it has a different pattern of enzyme functions from the cytoplasmic membrane (see Table 3.3); it also appears to be relatively free of particles in freeze-fracture replicas or in negative-stained preparations (see p. 8). Nevertheless, there are good reasons to believe that in another sense mesosomes are artefacts. If, instead of fixing and staining organisms as above, cultures of rapidly growing *Streptococcus faecalis* are frozen immediately in liquid nitrogen and examined by the freeze-fracture technique no mesosomes are seen. They appear if the culture is cooled slowly and centrifuged or if the bacteria are either treated with glutaraldehyde or fixed by osmium tetraoxide using the standard techniques. It is to be noted that the mesosomes still appear in their customary situation, that is associated with the division sites of the bacteria. If when the bacteria are treated with glutaradehyde a radioactively labelled amino acid is added simultaneously, the rate at which it was fixed to the preparations corresponds to that for the appearance of mesosomes. It is thus tempting to suggest that the mesosomes appear as a consequence of the cross-linking reaction carried out by glutaraldehyde. Perhaps the best hypothesis is that they are formed as a result of insults to the bacteria that set up some type of cross-linking reaction. However, this must occur in a special region of the cytoplasmic membrane which is in some (as yet unknown) way specificially differentiated to allow cell division to occur, and which agglomerates as mesosomes.

Before leaving the subject of mesosomes it may be well to discuss some of the major hypotheses that have been promoted as to their function and to dismiss some of the most important ones. Mesosomes cannot be mitochondrial equivalents as was once suggested, because they lack or have very low levels of essential components of the electron transport chain such as dehydrogenases and most of the cytochromes (see Table 3.3). For the same reasons they are not concerned with wall synthesis. Any relation to membrane lipid synthesis seems unlikely in view of the results of pulse-chase experiments designed to demonstrate their biosynthesis. Questions, however, still hang over their possible roles in cell division, export of proteins, and in transformation with exogenous DNA. Whatever their function in the living growing cell they are likely to exist as membranous material capable of coalescing to give visible mesosomes rather than as the familiar structures seen in prepared sections of bacteria.

The outer membrane of Gram negative species

The outer membrane appears in section as a wavy, characteristically double tracked membrane of a thickness varying from 6 to 10 nm according to the species. In some circumstances it can be made to form blebs (e.g. growth limitation by lysine of a lysine auxotroph of *E. coli*, or in rough strains of *Pseudomonas aeruginosa*). Lipopolysaccharide and other outer membrane constituents have even been found in the culture supernatants of exponentially growing wild-type strains of organisms such as *Salmonella typhimurium* or *Neisseria meningitis*, suggesting that rapidly growing organisms may pinch off and shed parts of that outer membrane. If Gram negative bacteria are negatively stained, pictures consistent with extremely corrugated surfaces for the organisms are obtained. It is probable, however, that this extreme corrugation is a reflection of the shrinkage of the cell that is known to take place during negative staining. When examined by the freeze-etching technique the outer surface of *E.coli* does not appear corrugated and even the waviness seen in section is not particularly apparent but rather the surface is smooth with scattered lumps on it. The way in which the outer membrane is thought to be linked with the peptidoglycan, by a lipoprotein covalently attached to the latter, has already been described (see pp. 10).

Much attention has been given to the isolation of the outer membranes particularly from *E. coli* and *S. typhimurium*. These methods have often been similar to that already described (p. 32) for the separation of the cytoplasmic membrane. However, methods specific for the outer membrane exist that depend upon characteristics other than its density. For example, neutral detergents can be applied to mixtures of cytoplasmic and outer membranes. Under the right conditions the cytoplasmic membrane can be dispersed whereas the outer membrane remains intact and can be obtained by centrifuging the mixture.

Membranes during sporulation

Among the early events in sporulation of bacilli is the formation of a septum consisting of a double membrane, usually occurring much nearer to one pole of the rods than to the other. Again a mesosome can be seen to be associated with this abnormal division, in stained and fixed bacteria. The membranous septum is then drawn out towards the distal pole (see Fig. 3.9) as sporulation proceeds. It is finally pinched off in the region of the proximal pole to become a forespore within the mother cell. The orientation of the forespore membranes is therefore of great interest since, as can be seen from Fig. 3.9, the two membranes should have opposite orientation. The outer membrane of the forespore should have the opposit orientation to that of the cytoplasmic membrane, whereas the inner membrane should have the same orientation. As would be expected from this predicted topology, enzymes using membrane-impermeant substrates like ATP and NADH cannot be detected in whole protoplasts or sporangia but can be readily detected in intact forespores. The reversed polarity of these membranes has fascinating implications for the transport of metabolites between the cytoplasm of the sporangium and the forespore.

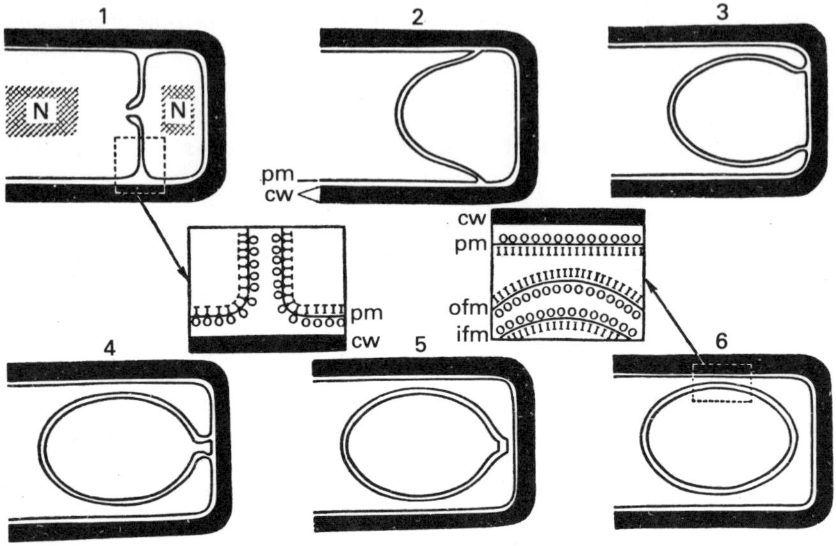

Fig. 3.9 The behaviour of membranes during the sporulation of bacilli during stages I to III. The total process of sporulation has six stages. The inset figures show the orientation or sidedness of the forespore membranes. PM = plasma or cytoplasmic membrane, CW = cell wall, OFM = outer forespore membrane, IFM = inner forespore membrane. (From Ellar D.J., 1978)

For example, the active transport of Ca^{2+} that is switched on during the formation of spores occurs only in the vegetative membrane. Movement of the ion from the vegetative cytoplasm into the forespore is a passive process not influenced by the presence of metabolic poisons. It is also into the space between the two forespore membranes that the primordial cell wall and the cortical peptidoglycan are secreted. Whether both inner and outer forespore membranes are active in this process is not known. Evidence from *B. sphaericus*, in which the cortical and vegetative wall peptidoglycans can be distinguished by an amino acid difference, shows that the formation of the primordial cell wall occurs first, the specific cortical peptidoglycan is only deposited after this. It may be that the inner and outer forespore membranes undertake separate biosynthetic tasks, though no firm evidence at present exists.

Specialized membranes

Among the micro-organisms that produce the greatest variety of morphological arrays of internal specialized membranes are photosynthetic bacteria. Fig. 3.10 shows the constellation of arrangements that have been described in different species and genera. Those that can grow in the dark as well as using photosynthetic energy switch on the formation of their internal membranes when exposed to light, making a very useful system for studying membrane morphogenesis and differentiation. They are clearly derived by infolding from the cytoplasmic membrane in the purple bacteria (Rhodospirillaceae), and the Chromatiaceae and bear a number of enzymes characteristic of this structure, such as succinic dehydrogenase. In the green sulphur bacteria, the Chlorobiaceae, chlorobium vesicles attached to the cytoplasmic membrane contain bacteriochlorophyll *c*, *d* or *e*, but the latter itself also contains chlorophyll a_p. Likewise in the purple bacteria, chlorophylls are not confined to the internal membranes but also occur in the cytoplasmic membrane. It will be noted, however, that although all the membranes contain some form or forms of chlorophyll the actual types in the cytoplasmic and specialized membrane are often different.

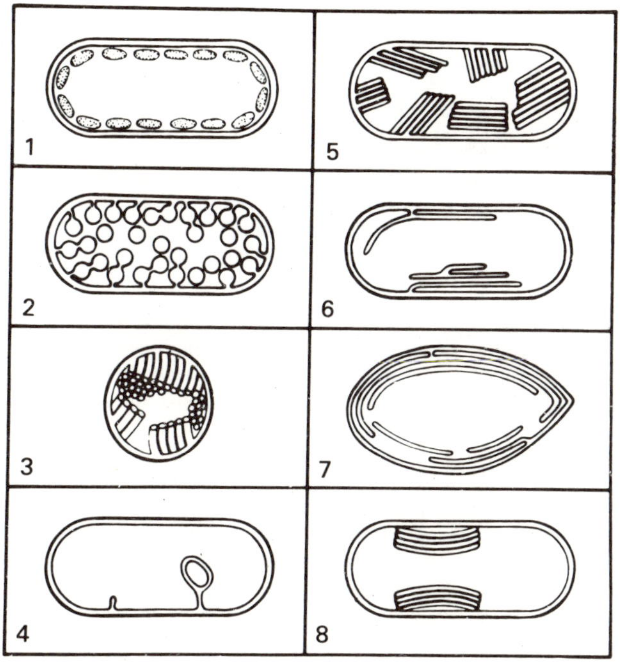

Fig. 3.10 Arrangements of intracytoplasmic membrane structures (chromatophores or thylakoids) of photosynthetic bacteria.

1 Chlorobium vesicles in the green sulphur bacteria *Chlorobium limicola, C. thiosulphatophilum, C. Phaeobacteroides, C. Phaeovibrioides, Pelodictyon clathratiforme, P. aggregatum, Chloropseudomonas ethylicum* and *Prothecochloris aestuarii*.

2 Vesicular membrane system in *Rhodospirillum rubrum, R. spheroides, R. capsulatei, Chromatium* strain D, *C. okenii, C. weissei, Thiospirillum jenense* and *T roseopersicina*.

3 Tubular membrane systems in *Thiocapsa pfennigii* (*Thiococcus sp.*).

4 Single small membrane invaginations in *Rhodopseudomonas gelatinosa* and *Rhodospirillum tenue*.

5 Stacks of short double membranes bound to the cytoplasmic membrane in *Rhodospirillum molischianum, R. fulvum, Ectothiorhodospira* and *R. photometricum*.

6 Parallel lamellae underlying and continuous with the cytoplasmic membrane in *Rhodopseudomonas palustris* and *R. acidophila*.

7 Concentric layers of double membranes in cells of *Rhodomicrobium vannielii*.

8 Grana-like stacks of double membrane in *Rhodopseudomonas viridis*.

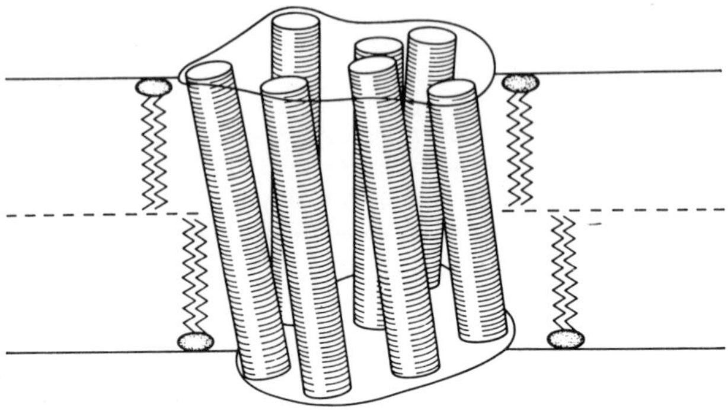

Fig. 3.11 The suggested arrangement of the protein α -helices in one protein molecule from the purple membrane of *Halobacterium halobium*. They are shown immersed in the lipid bilayer of the membrane. (from Henderson, 1978)

The purple membrane of halobacteria is so called because of the colour given by a carotenoid pigment very similar to rhodopsin, the mammalian visual pigment. It occurs in patches in the cytoplasmic membrane of *Halobacterium halobium* and is highly organized with a clearly recognizable hexagonal lattice in freeze-fracture replicas. The protein which makes up 75% of the weight of the membrane has a molecular weight of 26,000 daltons and has one molecule of retinal per molecule of protein, to which it is bound via a Schiff base to a lysine residue. The protein molecules are each composed of seven α-helical segments about 4 nm long which traverse the membrane. Fig. 3.11 shows the suggested arrangement of one molecule of bacterial rhodopsin. This result was obtained from high resolution (0.7nm) low dosage electron microscopy. Purple membrane can act as a proton pump using photochemical energy, thus creating an electrochemical gradient across the membrane.

'Pseudomembranes'

A variety of particles and vacuoles have been described in the cytoplasm of bacteria. Many of these are encased within membrane-like structures. Where it has proved possible to isolate these sacs the membranes have been shown to have a very different composition from the cytoplasmic and other 'true' membranes. They are usually much thinner and frequently made mostly of protein; they do not show the characteristic bilayer appearance in section. The material encasing gas vacuoles in the cyanobacteria has been particularly carefully examined. These vacuoles are about 70 nm in diameter and from 400 to 1000 nm long. By the freeze-fracture technique they can be seen to be ribbed on both sides of the membrane. The ribs, which are about 4.5 nm wide, run as a shallow spiral normal to the long axis. Chemical examination of such pseudomembranes isolated from not only cyanobacteria but also from halobacteria and another unidentified halotrophic bacterium, showed them to consist only of protein. Further work on the structures from cyanobacteria showed only a single highly hydrophobic protein to be present with a molecular weight of about 15,000 daltons, although one claim for the higher value of 50,000 daltons has been made. It is a sobering thought that this protein accounts for about 3.5% of the total protein in *Microcystis aeruginosa* and forms one of the ten most abundant proteins in *Anabaena flos-aqueae*. This may give some indication of the evolutionary importance of the gas vacuoles in making these microbes buoyant in their environment.

Composition of bacterial membranes

In common with membranes from other sources, bacterial membranes consist of a mixture of phospholipids and proteins although the proportion of the latter is higher in bacteria than in most membranes from eukaryotes (Table 3.4). The proteins present can be differentiated, again as in other membranes, into the two classes already referred to, namely the integral and peripheral proteins. The ATPases are a classical example of peripheral enzymes whilst integral proteins are well represented by succinic dehydrogenase and some other dehydrogenases, as well as by the cytochromes. The remainder of this chapter will be devoted to a summary of known chemical characteristics of bacterial membranes that differentiate them from those of other sources.

Table 3.4 Chemical composition of membranes from Gram positive bacteria and micro-fungi

Organism	Protein (%)	Total lipid (%)	Hexose (%)	RNA (%)
Bacillus megaterium	58–75	20–28	0.2–8	1.2–5.1
Micrococcus luteus	52–68	23–28	16–19	2.3
Staphylococcus aureus	69–73	30	1.7	2.4
Saccharomyces cerevisiae	46–47.5	37.8–45.6	3.2	6.7
Candida utilis	38.5	40.4	5.2	1.1
Candida albicans (mycelial form)	45.0	31.0	25	0.5
Fusarium culmorum	25.0	40.0	30.0	—

The absence of sterols Membranes from mammals, plants and eukaryotic microorganisms such as yeasts always contain sterols, cholesterol being a common component. Yeasts, for example, contain a very wide variety of different sterols. The bacteria that have been examined, on the other hand, contain no detectable amounts of sterols except the Acholeplasmae that have no walls. However, as was said earlier, they may perhaps be regarded only as honorary bacteria. The exact significance of the absence of sterols from bacterial membranes is not clear. It is thought that their presence in other membranes affects the fluidity and therefore rates of movement of the other components and the effect of temperature changes upon the fluidity. Evidence in favour of this has come from, among other sources, the study of certain *Acholeplasma* membranes in which the sterol content can easily be varied.

Fig. 3.12 Structures of phospholipids and fatty acids present in bacterial membranes

(a)

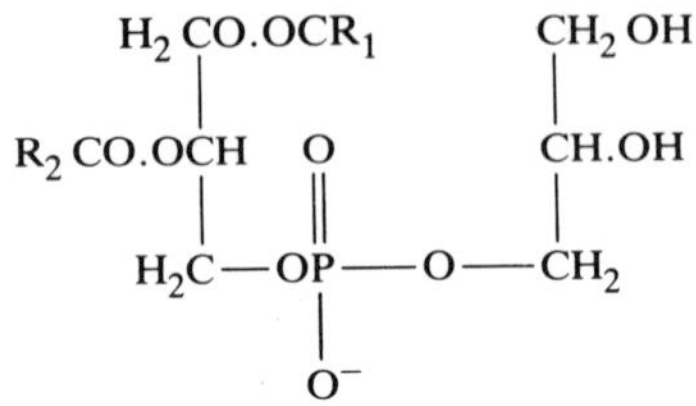

Phosphatidylglycerol

(b)

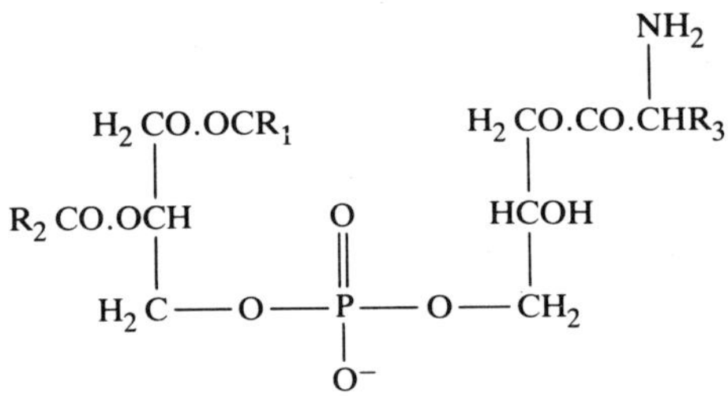

Amino acyl ester of phosphatidylglycerol

(c)

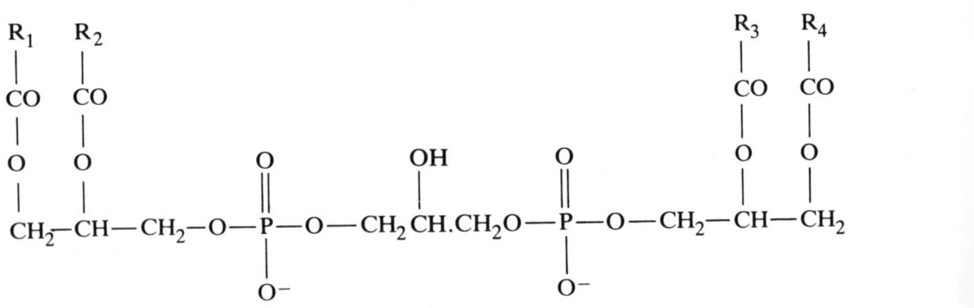

Phosphatidylglycerol (cardiolipin)

(d)

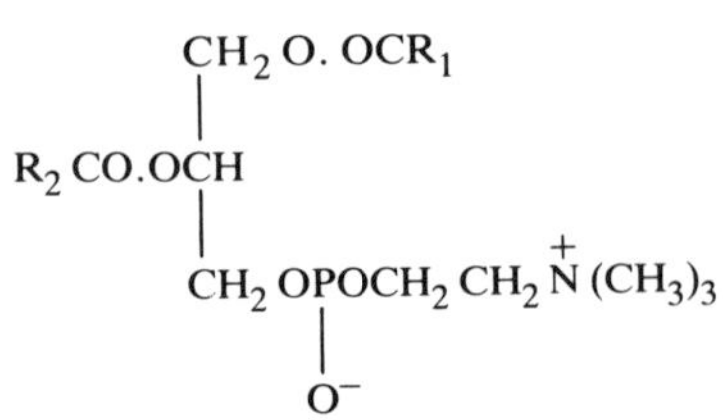

Phosphatidylcholine (lecithin)

(e)

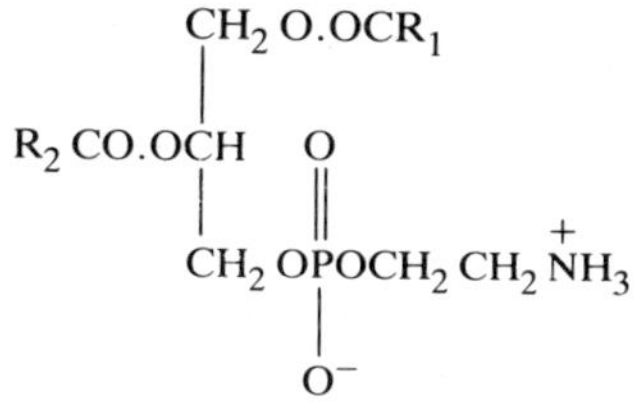

Phosphatidylethanolamine

(f)

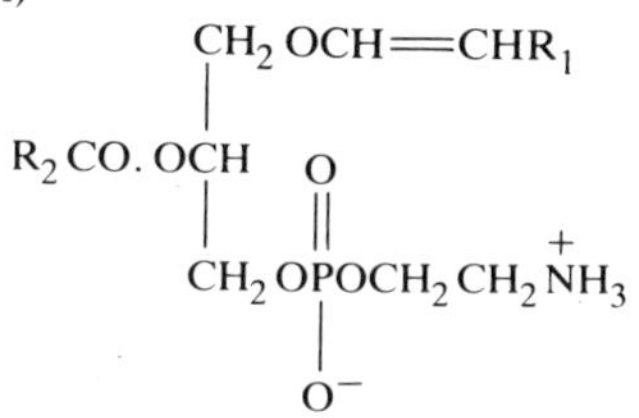

Phosphatidal ethanolamine

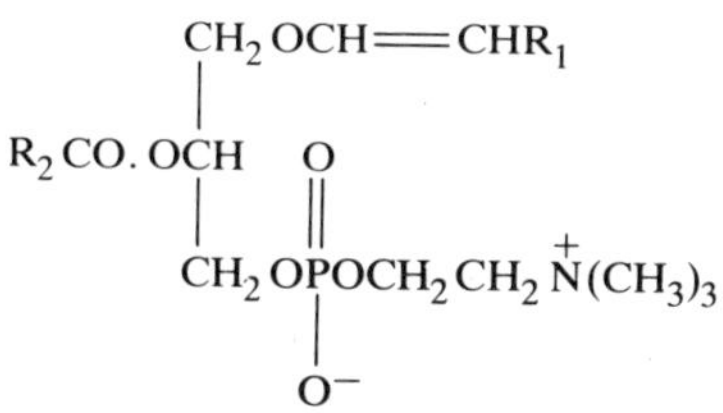

Phosphatidal choline

Plasmalogens

(g)

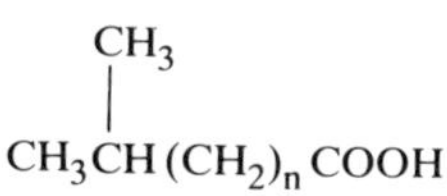

iso fatty acid

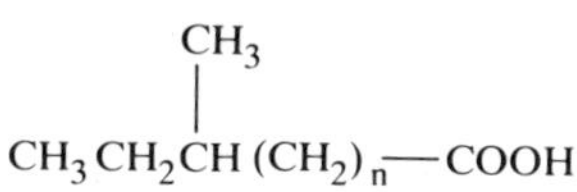

antiiso fatty acid

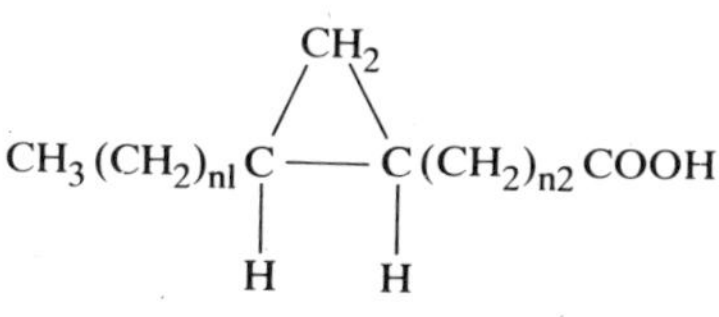

cyclopropane fatty acid

Branched chain and cyclopropane fatty acids

(h)

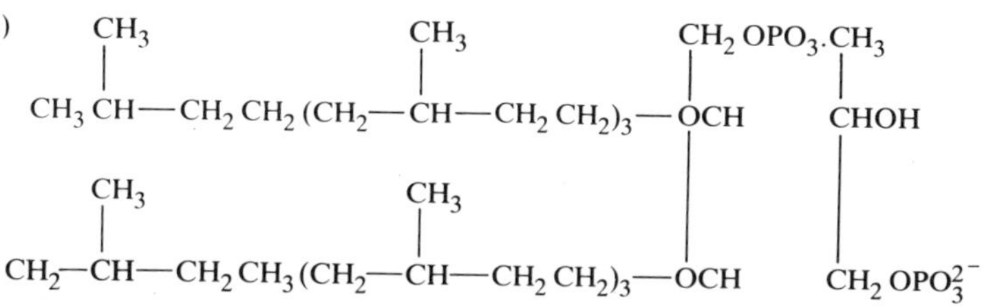

Structure of the dihydrophytyl ether analogue of phosphatidylglycerol phosphate present in extreme halophiles

Table 3.5 The glycosyl diglycerides of known structure in bacteria

Carbohydrate residues	Organism
Monoglycosyl	
Glcα1 →	*Acholeplasma laidlawii*
	Acholeplasma modicum
Glcβ1 →	*Mycoplasma neurolyticum*
Galα1 →	*Treponema pallidum*
Galβ1 →	*Arthrobacter sp.*
Galfβ1 →	*Mycoplasma mycoides*
	Bifidobacterium bifidum
GlcAα1 →	*Pseudomonas diminuta*
GlcAβ1 →	*Pseudomonas nibexcus*
GlcNβ1 →	*Bacillus megaterium*
Diglycosyl	
Glcα1 → 2Glcα1 →	*Streptococcus lactis*
	Streptococcus haemolyticus
	Streptococcus faecalis
	Acholeplasma laidlawii
Glcβ1 → 6Glcβ1 →	*Staphyloccus lactis*
	Staphylococcus saprophyticus
	Staphylococcus aureus
	Bacillus subtilis
	Mycoplasma neurolyticum
	Pseudomonas iodinium
Galβ1 → 6Galβ1 →	*Arthrobacter crystallopoietes*
	Arthrobacter pasceus
	Arthrobacter globiformis
Galβ1 → 2Galβ1 →	*Bifidobacterium bifidum*
Galβ1 → 2Galfβ1 →	*Bifidobacterium bifidum*
Manα1 → 3Manα1 →	*Micrococcus luteus*
	Arthrobacter sp. (see above)
Galα1 → 2Glcα1	*Diplococcus pneumococcus* type 1
	Diplococcus pneumococcus type XIV
	Lactobacillus casei
	Lactobacillus buchneri
	Lactobacillus plantarum
	Lactobacillus helveticus
	Lactobacillus acidophilus
	Listeria monocytogenes
Glcβ1 → 4GlcAα1	*Pseudomonas diminuta*
Glcα1 → 4GlcAα1	*Streptomyces* LA 7017
Triglycosyl	
Glcα1 → 2Glcα1 → 2Glcα1 →	*Streptococcus haemolyticus*
Glcα1 → 6Galβ1 → 2Glcα →	*Lactobacillus casei*
	Lactobacillus helveticus
	Lactobacillus acidophilus
3′-SO_3-Galβ1 → 6Manα1 → 2Glcα1 →	*Halobacterium cutirubrum*
Galβ1 → 2Galβ1 → 2Galβ1 →	*Bifidobacterium bifidum*
Tetraglycosyl	
Glc1 → 6Glcα1 → 6Galα1 → 2Glcα1 →	*Lactobacillus acidophilus*
Galf1 → 2Gal1 → 6GlcNHAc1 → Glc1 →	*Flavobacterium thermophilum*

Types of phospholipid Membranes from many bacteria are distinguished by the absence or very small content of sphingomyelins and phosphatidylcholine (lecithin), the latter of which is a major component of membranes from other sources. Some phosphatidylcholine, however, has been found in species of *Agarobacter*, *Nitrocystis*, *Thiobacillus* and *Ferrobacillus*. On the other hand bacterial membranes, particularly from the cocci, contain very high concentrations of phosphatidyglycerol and diphosphatidylglycerol which are often amino-acylated, lysine and alanine commonly being involved. The degree of amino-acylation, like the overall proportion of phospholipids, is often strongly influenced by the conditions for growth of the bacteria (see Fig. 3.12 for structure of phospholipids).

Fatty acid composition of the phospholipids Whereas membranes from other sources commonly contain polyunsaturated fatty acids (e.g. oleic or elaidic acids) but do not contain branched chain or cyclopropane fatty acids (see Fig. 3.12), the reverse is true for bacteria. Polyunsaturated acids are found only in a few bacteria including *Mycobacterium phlei*, whilst branched chain iso- and anteiso fatty acids are very commonly present. Again the growth conditions for the micro-organisms can have a profound effect on the fatty acid composition – one common example is the accumulation of cyclopropane fatty acids when growth is limited by the supply of oxygen. It should be mentioned that the mycobacteria, which in so many other respects represent departures from generalizations about bacterial envelopes, have highly complex unsaturated and cyclopropane fatty acids (the mycolic acids) as components of their cell walls (see Chapter 2 Table 2.8). Compounds with a similar general structure have also been found in the corynebacteria.

Lipids containing carbohydrate Bacterial membranes are a particularly rich source of a large variety of sugar-containing lipids. Among them are the glycolipids and the glycophospholipids. A wide variety of glycolipids has been found. These are 1, 2 diacyl-*sn*-glycerol with carbohydrates linked to the 3-hydroxyl group. The structures of these compounds are summarized in Table 3.5. The glycophospholipids are similar to the glycolipids except that the carbohydrate is linked to the 3-hydroxy group of the glyceride residue through a phosphodiester group rather than directly so. Among the earliest to be discovered were a group of di-, tri-, tetra- and pentamannosyl derivatives of inositol isolated from mycobacteria (see Fig. 3.13). Since that time they have been recognized in many other organisms such as species of actinomycetes, *Nocardia* and corynebacteria, *Streptococcus*, *Bacillus*, *Listeria* and *Pseudomonas*. Their location is not strictly established except in some bacilli, but it is most probable that they also have a membrane origin in organisms such as *Streptococcus faecalis*, *Listeria monocytogenes* and pseudomonads; this is less certain in the remainder of the species mentioned.

Anisotropy of membranes It is by now generally accepted that the lipid composition of the two bilayers making up membranes is different. This is particularly true for the outer membranes in Gram negative bacteria, providing results obtained with a very limited number of species can be generalized. It would appear that the normal phospholipids, of which, in the examples examined, phosphatidylethanolamine forms a high proportion, are confined to the inner leaflet of the bilayer, whilst in the outer leaflet the lipid A of lipopolysaccharide dominates the picture.

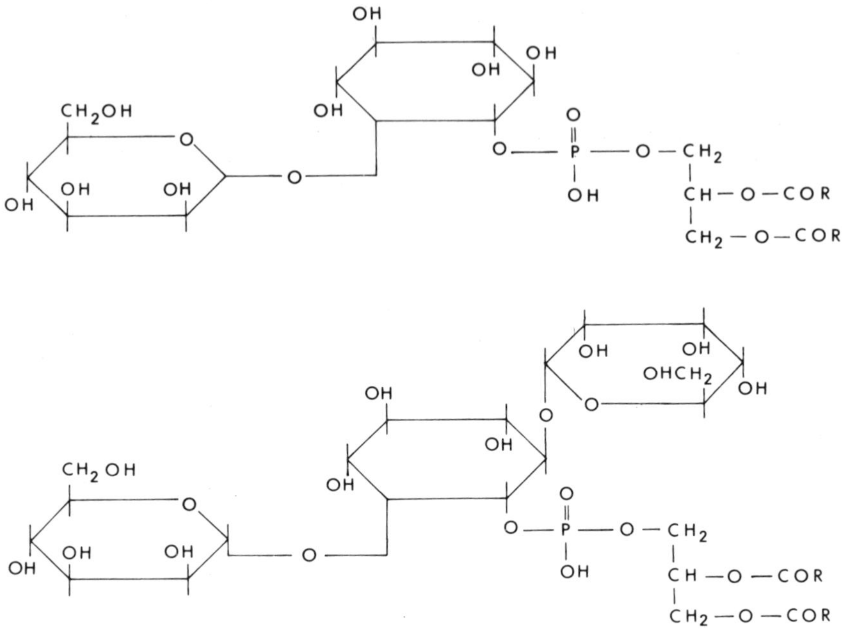

Fig. 3.13 Structures of the phosphatidylinositol mono- and dimannoside isolated from mycobacteria. R represents fatty acid residues.

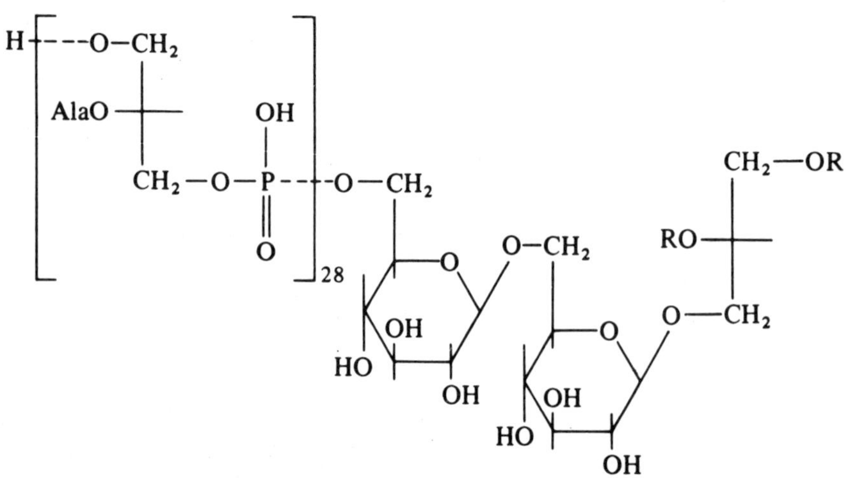

Fig. 3.14 The structure of a lipoteichoic acid molecule. R represents fatty acid residues.

Lipoteichoic acids and lipomannans The lipoteichoic acids, formerly known as membrane teichoic acids, are, like some of the wall teichoic acids, substituted polyglycerol phosphates. They are not covalently attached to the wall but terminate with a glycolipid (see Fig. 3.14) presumably buried in the cytoplasmic membrane. They are very common among Gram positive organisms and differ in their glycoside substituents. It would appear that in some organisms such as lactobacilli the hydrophilic parts of the lipoteichoic acid molecules extend throughout the wall, so that antibodies specifically directed against them will react with whole cells. In other organisms it has been shown that lipoteichoic acid, often in a deacylated form, is exported through the wall to reach the surrounding culture fluid. The location of the lipoteichoic acids with one end in the cytoplasmic membrane and the other extending through the wall suggests that they may act to keep the wall and membrane in communication. However, many functions have been proposed for them with suggestive but not conclusive evidence in their favour. The lipoteichoic acids strongly inhibit some autolysins such as the muramidase of *S. faecalis* and *Lactobacillus acidophilus* and the β-*N*-acetylglucosaminidase of *Bacillus subtilis*. The *N*-acetylmuramyl-L-alanine amidase of *B. subtilis* is inhibited by lipoteichoic acid from *S. faecalis* but not by the homologous compound. The nature of the substituents on the polyglycerolphosphate chain did not seem to be important. The Forssman antigen from pneumococci (which is probably a choline containing lipoteichoic acid or an analogous compound) strongly inhibits pneumococcal autolysin, which is an amidase, but fails to inhibit the muramidase of *S. faecalis*. Deacylated lipoteichoic acid has no inhibitory effect.

It has been suggested that lipoteichoic acids are important in regulating the action of the autolysins and that the autolysins are involved in the bacteriocidal action of antibiotics, such as β-lactams (e.g. penicillins). Consequently the lipoteichoic acids could be important in the action of β-lactam antibiotics. In support of this it has been claimed that the antibiotics cause rapid excretion of lipoteichoic acid and other membrane phospholipids. Addition of lipoteichoic acid to cultures of staphylococci treated with β-lactam antibiotics prevented death of the micro-organisms. The lipoteichoic acids have also been implicated in the ability of organisms such as streptococci to adhere to the surface of mammalian cells. The lipophilic end of the molecule appears to interact with the mammalian cell surface and the precise mechanism is unclear. It has also been suggested that the lipoteichoic acids are important in regulating the supply of Mg^{2+} to the membrane enzymes. Clearly much work is yet necessary before these various hypotheses can be tested so as to give unambiguous answers, not least to distinguish clearly the specific effects of lipoteichoic acid from those of other amphipathic molecules.

As was implied the presence of lipoteichoic acid is almost universally present among Gram positive organisms. A few, however, such as *Micrococcus luteus*, *M. flavus* and *M. sodenensis* lack the compound. However they have a lipomannan instead. This compound is an α-linked branched mannan of 50 to 70 residues linked glycosidically to a diglyceride, which is presumably part of the membrane. About 25% of the mannose residues have succinyl 0-acyl substituents: the free —COOH groups of these residues give the mannan a strongly negative charge. The functions of the mannan have been examined but the analogy with the lipoteichoic acids is clear and tempting.

Protein composition of the outer membrane of Gram negative species The protein composition of the outer membrane differs from that of orthodox membranes

Table 3.6 Recommended genetic nomenclature for major outer membrane proteins of *E. coli* and *S typhimurium* (from Osborn and Wu, 1980)

Gene	Map position (units)			Protein nomenclature								Recommended uniform nomenclature
				Previous designations *E. coli*								
	E. coli	*S. typhimurium*	Gene† function	B	F	H	I	L	M	S	*S. typhimurium*	
ompA (*tolG, con, tut*)	21	21	S	B	TolG	11*	7	D	0-11	3a	33K	OmpA
ompB (*cry*)	74	74	R									
ompC (*par, meoA*)	47	46	S	A_2		1b	4		0-8	1b	36K	OmpC
ompD	—	28	S(?)								34K	OmpD
ompF (*folF, colB, coa, cry*)	21	21	S	A_1	TolF	1a	4		0-9	1a	35K	OmpF
rmpA	82.2	—	?		E	1c				NmpA		NmpA-B
nmpB	8.6	—	?							NmpB		
nmpC	12	—	?							NmpC		NmpC
lamB	90		S								44K	LamB
Ipp (*mlpA, Ipm, lpo*)	36.5		S	F		IV	11		0-18			Murein lipoprotein

† S–structural, R–regulatory, ?–not known.

(i.e. those having the double track appearance in section see p. 00) which include the cytoplasmic membrane from bacteria. When components of orthodox membranes are separated by unidirectional polyacrylamide gel electrophoresis, after dispersion in anionic detergents such as sodium dodecyl sulphate, 30 to 40 bands are resolved which stain to different intensities with reagents such as Coomassie Blue (which reacts with proteins or polypeptides). The rates of movement through the gels suggest molecular weights for these proteins ranging from about 1500 daltons to several 100,000 daltons. This technique for various reasons may provide a very conservative estimate of the total number of proteins present. Indeed, if they are first separated in one direction by iso-electric focusing and then in a second direction by gel electrophoresis, (i.e. two dimensional electrophoresis) 200 or more spots corresponding to proteins and polypeptides are seen. Moreover membranes, especially the bacterial cytoplasmic membrane, demonstrate a wide range of enzymic activities as has already been discussed.

The protein composition of the outer membranes from a considerable number if not all Gram negative species differs strikingly from this general picture. Only two enzymic activities, a protease and a phosphatase, have been detected in these structures. Although a large number of proteins can be seen in polyacrylamide gels, four or five are present in overwhelmingly large amounts. Indeed, unless rather large amounts of the dispersed protein preparations are applied it appears that these are the only proteins present. Unfortunately for those interested in this subject the nomenclature of these few proteins is highiy confusing and some seven or eight different names have been given to them (see Table 3.6). One of them, the smallest, is the lipoprotein which is also found covalently attached to the peptidoglycan (see p. 10). Two (or in some species only one) of the others have been named porins because they are transmembrane proteins and there is strong evidence that they interact so as to form aqueous pores through the membrane. In the Enterobacteriaceae the

Table 3.7 Properties of some of the proteins present in lesser amounts in the outer membranes of Gram negative species (modified from Dirienzo *et al.*, 1978)

Gene specifying protein	Molecular weight (daltons × 10^{-3})	Uptake function	Bacteriophage receptor
feuB	81	Fe^{3+}-enterochelin	ColB
cit	80.5	Fe^{3+}-citrate	—
tonB	78	Fe^{3+}-ferrichrome	T1, T5, ∅80 Albomycin Colm
cir[a]	74	—	Col1, ColV
fbe	60	Vitamin B12	BF23, ColE1 ColE2, ColE3
lamB	55	Maltose	λ
tsx	27	Nucleosides	T6, ColK

Represented by Fe^{3+}

pore diameter is sufficient to exclude hydrophilic molecules with molecular weights greater than about 600. Some evidence suggests that these pores may open and close thus behaving rather like stomata on green leaves: this however, remains to be confirmed. In other species such as the pseudomonads it is probable that the pores are a good deal wider with exclusion limits of several thousands. Another of the proteins has been called TolG by one system of nomenclature and *omp* A by another, and may have a function in F-pilus mediated sex conjugation between cells.

Many of the important functions of the outer membrane are carried out by proteins usually present in much smaller amounts than the four or five mentioned above and these can only be detected with difficulty in gels. Some of them have multiple functions such as acting as bacteriophage and/or colicin receptors as well as in specifically facilitating the passage of molecules such as iron-chelate compounds through the outer membranes. Their properties are summarized in Table 3.7. The outer membrane also excludes hydrophobic compounds from the cells. Less is known about the selectivity of this system but the polysaccharide chains of the lipopolysaccharide are likely to be involved since deep rough mutants lacking these chains are less effective in excluding toxic hydrophobic substances. For this to happen not less than 80 to 90% of the polysaccharide must be missing. Morever, in such mutants some of the outer membrane proteins are also absent and more work is necessary to resolve the relative roles of the two deficiences.

Summary

The arrangement of membranes in many bacteria is very simple. A cytoplasmic membrane is closely applied to the wall. Foldings giving rise to mesosomes can be seen in the region of cell division and sometimes elsewhere in cells prepared for electron microscopy by standard techniques. Mesosomes cannot be seen if rapidly growing bacteria are quickly frozen in liquid nitrogen and examined by freeze-fracture without fixing or staining. In Gram negative bacteria a further outer membrane is present. The membranes can be separated and their chemistry has been examined. Some species of bacteria such as the photosynthesizers, nitrogen fixers, methanogenic and methane utilizing bacteria have ample internal membranes that occur in a variety of rather simple arrangements. Some of these membranes are likely to have their origin from the cytoplasmic membrane.

Chemically, bacterial membranes are distinguished by the following characteristics: (a) the absence of sterols; (b) the absence, in most species, of polyunsaturated fatty acids but the presence of branched chain anti and anteiso fatty acids, (c) the absence of sphingomyelins and the very common, although not universal, absence of phosphatidyl choline (lecithin), and in some bacteria the high content of phosphatidylglycerol and diphosphatidylglycerol, (d) the presence of a wide variety of sugar-containing lipids, particularly glycolipids. The outer membrane of Gram negative bacteria can be distinguished from other membranes including the cytoplasmic membranes by a number of factors. Firstly, anisotropy (i.e. the two layers of the lipid bilayer are not identical) is extreme, with normal phospholipids confined only to the inner leaflet of the bilayer, and their place being taken by the lipid A of lipopolysaccharide in the outer leaflet. Secondly, their protein composition is unique in that four or five proteins are present in very much greater amount than any others. Some of these have the function of forming pores that regulate the passage of hydrophilic substances. Thirdly, very little enzymic activity is present in outer membrane preparations.

References

DiRienzo, J. M., Nakamura, K. and Inouye, M. (1978). The outer membrane proteins of Gram-negative bacteria. Biosynthesis assembly and functions. *Annual Review of Biochemistry* 47: 481–532.

Downie, J. A., Gibson, E. and Cox, G. B. (1979). Membrane adenosine-triphosphatases of prokaryotic cells. *Annual Review of Biochemistry* 48: 103–31.

Ellar, D. J. (1978). Spore specific structures and their function. *Symposium of the Society for General Microbiology* 28: 295–325.

Goldfine, H. J. (1972). Comparative aspects of bacterial lipids. *Advances in Microbial Physiology* 8: 1–51.

Mahendra, K. J. and White, H. B. (1977). Long-range order in biomembranes. *Advances in Lipid Research* 15: 1–60.

Osborne, M. J. and Wu, H. C. P. (1980). Proteins of the outer membrane of Gram-negative bacteria. *Annual Review of Microbiology* 34: 349–422.

Rogers, H. J., Perkins, H. R. and Ward, J. (1980). In: *Microbial Cell Walls and Membranes*. Chapman and Hall, London and New York. Chapters entitled: Isolation of walls and membranes, Membrane structure and composition in micro-organisms, and Membrane functions.

Salton, M. R. J. (1978). Structure and function of bacterial plasma membranes. *Symposium of the Society for General Microbiology* 28: 201–23.

Salton, M. R. J. (1980). Structure-function relationships of *Micrococcus lysodeikticus* membranes. A bacterial membrane model system sub-cellular biochemistry. In: *Subcellular Biochemistry* Vol 7, pp. 309–73. Edited by D. B. Roodyn. Plenum Publishing Corporation.

Shaw, N. (1970). Bacterial glycolipids. *Bacteriological Reviews* 34: 365–77.

Shaw, N. (1975). Bacterial glycolipids and glycophospholipids. *Advances in Microbial Physiology* 12: 141–7.

4 The nuclear material and the cytoplasm

Of the morphologically recognizable parts of living bacteria we know least about the organization of the DNA and the cytoplasm; that is, the materials contained within the bounding cytoplasmic membrane. We know a lot about the structure and function of isolated ribosomes and about the structure and function of isolated DNA but very little about how either is organized within the living and growing cell. We know nothing about the organization of components such as enzymes found in the cell sap. The reasons for this ignorance are the close packing of materials in bacteria, together with the suspicion that almost all the techniques used to explore the cytoplasm cause such changes in organization that only cautious conclusions can be drawn. In some instances insufficient techniques are available. Many preparative techniques for exploring ultrastructure with the electron microscope involve fixing and staining with reactive chemicals, dehydrating until almost all the cell water is removed, embedding in resins, cutting sections and examining them in a high vacuum. Better from these points of view are those techniques that use freeze-dried material or fast frozen material in the freeze-fracture and freeze-etch methods. In both groups of methods high proportions of the bacteria are kept alive. For example over 80% of *E.coli* have been shown to retain viability, so that hopefully not too much disorganization of the machinery has occurred. Although invaluable for examining the arrangements of membranes, (which fracture through their fatty interiors) and other structural and superficial components of bacteria and spores, such as cell walls, rather little information about the cytoplasm has been gained by these methods. Too often depressingly little is seen unless the bacteria have been treated first with the noxious materials such as osmium tetroxide which probably cause distortion. Even the ribosomes are so closely packed together that only an impression can be gained: the DNA cannot be seen.

More information about the interrelations of the cytoplasmic components has been obtained by gently bursting sphaeroplasts and protoplasts made from bacteria such that their contents can spread onto a suitable surface. Consideration of the results from such techniques is better delayed until we come to deal with the bacterial nucleus or nucleoid and the ribosomes as such. Meanwhile transverse sections examined after fixing and staining the bacteria by orthodox methods give reproducible pictures (see Fig. 4.1) that have become accepted as the 'ultrastructure' of bacteria without strong evidence for its reality in the growing dividing micro-organism.

The bacterial nucleoid

One of the constant features distinguishing the prokaryotic bacteria from eukaryotic micro-organisms such as yeasts, *Chlamydomonas* or amoebae is the absence of a nuclear membrane. Nuclear bodies in bacteria were observed many years ago by ordinary light microscopy after application of the common cytological procedures such as Feulgen or Giemsa staining. Moreover they can be seen by phase contrast microscopy without staining if the bacteria are immersed in high concentrations of

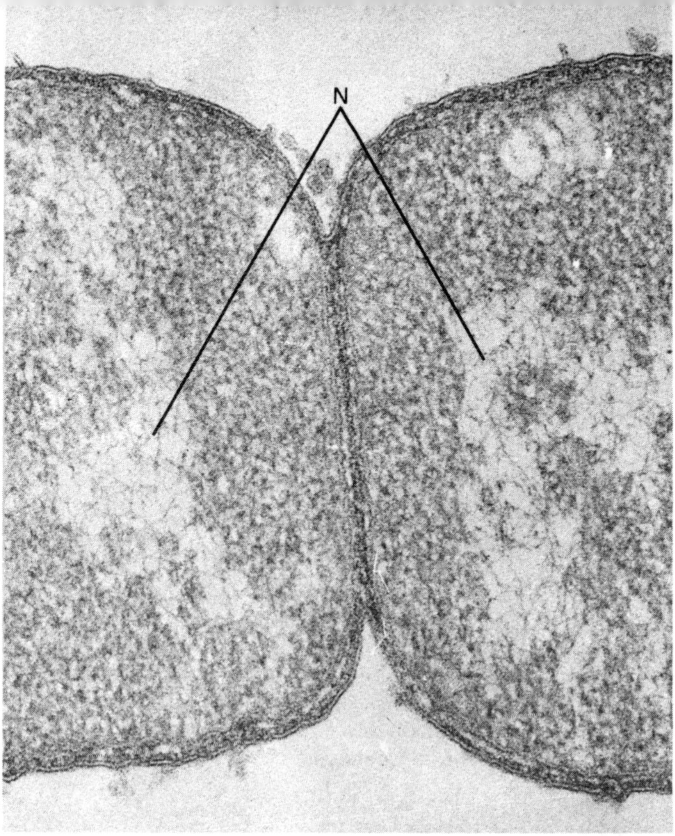

Fig. 4.1 A high powered view of part of a longitudinal section of *E. coli* cells that have just undergone division. The clearer areas (N) covered by fibres represent the nuclear material. (Photo: Dr Ian Burdett)

gelatin. Other methods involve the use of fluorescent stains such as acridine orange, ethidium bromide or DAP 1. Examination of thin sections of bacteria showed that indeed the nucleoid bodies were not surrounded by nuclear membranes such as are readily seen in eukaryotes. Genetic studies have shown that the chromosomal part of the 1.2×10^{-14}g of DNA contained in each exponentially growing *Escherichia coli* is organized as a continuous double-stranded circle replicating bidirectionally. By carefully bursting the bacteria and spreading out the DNA, Cairns was able to recognize molecules 1100 to 1400 μm long, some of which were closed circles and therefore likely to be a morphological realization of the circular chromosomes of the geneticist. After replication the new chromosome is segregated from the old and passes into the new cell. Two important problems posed to the morphologist by this information are how the long circular material is packaged in the living growing cell, and how the new and old chromosomes are segregated.

Areas of lesser electron density are seen in sections of bacteria fixed and stained by the Ryter-Kellenberg technique. These are traversed by skeins or bundles of fibres, thought to be the DNA (see Fig. 4.1). The exact appearance of the nuclear material in sections, however, depends upon a number of factors including the nature of the fixation and staining procedures. For example, increasing the ionic strength of fixatives leads to condensation of the nuclear material; treating with bifunctional aldehydes like glutaraldehyde leads to dispersed fine fibres. The appearance also depends on the strain of micro-organism. For example, in *B. subtilis* DNA is often seen as lighter areas around the mid-point of the bacteria with fibres visible, whereas in *E.coli* it is frequently much more dispersed and fibres are not so often seen. Again the form of the DNA can be altered by the growth medium and fixing procedures. Bacteria from rapidly growing cultures of *B. subtilis* in broth appear different from those from slowly growing cultures in minimal salts + glucose media. Although the DNA is usually most obvious in the central regions of an organism like *B.subtilis*, lighter areas are frequently scattered throughout the cytoplasm. Serial sections, however, show that these apparently separate areas are all linked to the central material. It has been suggested that the irregularity seen in the nuclear material may be because the transcriptionally more active areas are covered with ribosomes which appear more electron-dense after the staining procedures. It may be noted that the form of the nuclear material seen in stained sections of bacilli that are beginning to sporulate undergoes a rather dramatic change. Instead of being irregularly shaped and, in any one section, often appearing as islands throughout the cytoplasm, it forms a longitudinal rod-shaped axial structure. This event is sufficiently regular and dramatic to mark stage I of sporulation. Doubts, however, have been expressed as to how far this change is related to the expression of spore-specific genes. Since the form of the DNA in bacteria is dependent upon the osmolarity and ionic strength of the solutions used to fix the organisms, it has been suggested that the change of its form in sporulating cells may be related to alterations in the intra-cellular solute concentrations which occur during sporulation.

Studies, both by serial sections and by molecular biological methods, have strongly suggested that the chromosome is attached to the cytoplasmic membrane by its origin and terminus. In a Gram positive bacterium such as *Bacillus subtilis* the mesosomal membrane has been suggested by a number of electron microscopists as the point of attachment. Indeed, in one study strands of DNA appeared to penetrate to the inside of the mesosome. However, in view of the likelihood that mesosomes as seen in sections of bacteria are a product of the fixation procedure (see Chapter 3), any attachment of DNA can only be to the region of membrane so affected. Since this is universally in the central area of the cell, where division will take place, the observed DNA-mesosome association may still be important to our understanding of the mechanism of nuclear segregation. It is also of interest that in one careful freeze-fracture study of the gradual appearance of mesosomes during treatment of *S.faecalis* with glutaraldehyde, nuclear material became increasingly visible at the same time. The DNA, assuming that the fibres shown in the freeze-fracture replicas are DNA, appears closely associated with the mesosomes. Examination of the association of the DNA with the outer surface of *Escherichia coli* has suggested that components of the outer membrane, the peptidoglycan layer and the cytoplasmic membrane are all present at the point of attachment. This observation suggests that attachment corresponds to an area of adherence between the different layers of the bacterial envelopes similar to those demonstrated by plasmolysis of cells. Clearly the question of membrane-DNA attachment is a complex problem requiring further work. The situation is, however, even more complicated since multiple attachments of DNA to membrane may occur, albeit these are weaker than those formed by the origin and terminus of the chromosome. Indeed it has been suggested that random rather than specific sequences of DNA are responsible for the majority of attachments. It is important to know whether 'random attachment' occurs in the living organism. Two general mechanisms of specific attachment have been suggested. A unique protein has been described that is involved with attachment of the chromosome origin, and up to 20 associations have been suggested involving RNA polymerase, mRNA and ribosomes. An essential condition of the replicon hypothesis to explain segregation of DNA during cell growth and division is that the DNA should be attached to the envelope. According to this hypothesis the new and old chromosomes are moved apart by the expanding envelope during growth in length of rod-shaped organisms.

In view of the uncertainties of visualizing the nuclear material in whole bacteria, it is not surprising that the isolation of intact functional nucleoids should be an attractive idea, and in 1971 methods for doing this from *E.coli* were developed. The organisms were suspended in 20% sucrose and the peptidoglycan layer hydrolyzed by treatment with lysozyme in the presence of EDTA. The resulting sphaeroplasts were ruptured by treatment with non-ionic detergents such as deoxycholate or Brij-58 in the presence of 1 M NaCl. Nucleoids still associated with the envelope could be separated by centrifuging through sucrose gradients. Nucleoids not associated with the cell envelope (free-nucleoids) have also been isolated by treating the bacteria for longer times with lysozyme and the resulting sphaeroplasts with ionic detergents such as Sarkosyl. Chemical analysis shows that the former type of preparation, as might be expected, contains much besides DNA and RNA; protein and peptidoglycan are also present. The 'free nucleoids', however, contain few components other than DNA, RNA and RNA polymerase, the latter forming a major part of the protein present. The RNA amounts to 30% by weight of the preparations. Only 4% of this is ribosomal RNA.

The form of the DNA present in isolated nucleoids has been studied both by 'free solution' methods such as gradient centrifugation, and by spreading on to the surface of a variety of solutions followed by floating on to suitably treated electron microscope grids. It is supercoiled and the electron microscope pictures show (see Fig. 4.2) the presence of a large number of loops irrespective of whether envelope-associated or naked nucleoid preparations are studied. The contour length of the longest loop seen in preparations from envelope-free material is about 20 nm and the number of loops has been estimated to be 140. More electron dense material has been found at the centre of such preparations and, on the basis of a reduction in density by treatment with RNAase, it is suggested that it represents RNA. Extensive treatment with the enzyme caused a shift in density of isolated nucleoids from 1600 S to between 4 to 500 S and an increase in viscosity. High ionic strengths or the presence of basic material such as polyamines is necessary to prevent disintegration of nucleoids. It will be remembered that the degree of condensation of the nuclear material, as seen in sections of bacteria, is related to the ionic strength of the fixative solutions used. Much of the discussion about the mechanisms of organization of the nucleoids is speculative. Models have been based on the idea that RNA holds together the separate supercoiled loops of DNA. Other models include RNA-RNA as well as RNA-DNA interactions. However, the 'reality' of the nucleoids as representative of the situation in living bacteria has been questioned, just as has the appearance of DNA when seen in fixed and stained sections of bacteria. In the presence of the high salt concentrations necessary to stabilize the nucleoids and of detergents used to prepare them, it is argued that the presence of the RNA at the centre of nucleoids (apparently stabilizing the DNA) could be the result of aggregation of nascent RNA chains, mRNA and rRNA and DNA-RNA molecules, all of which are known to be possible under the

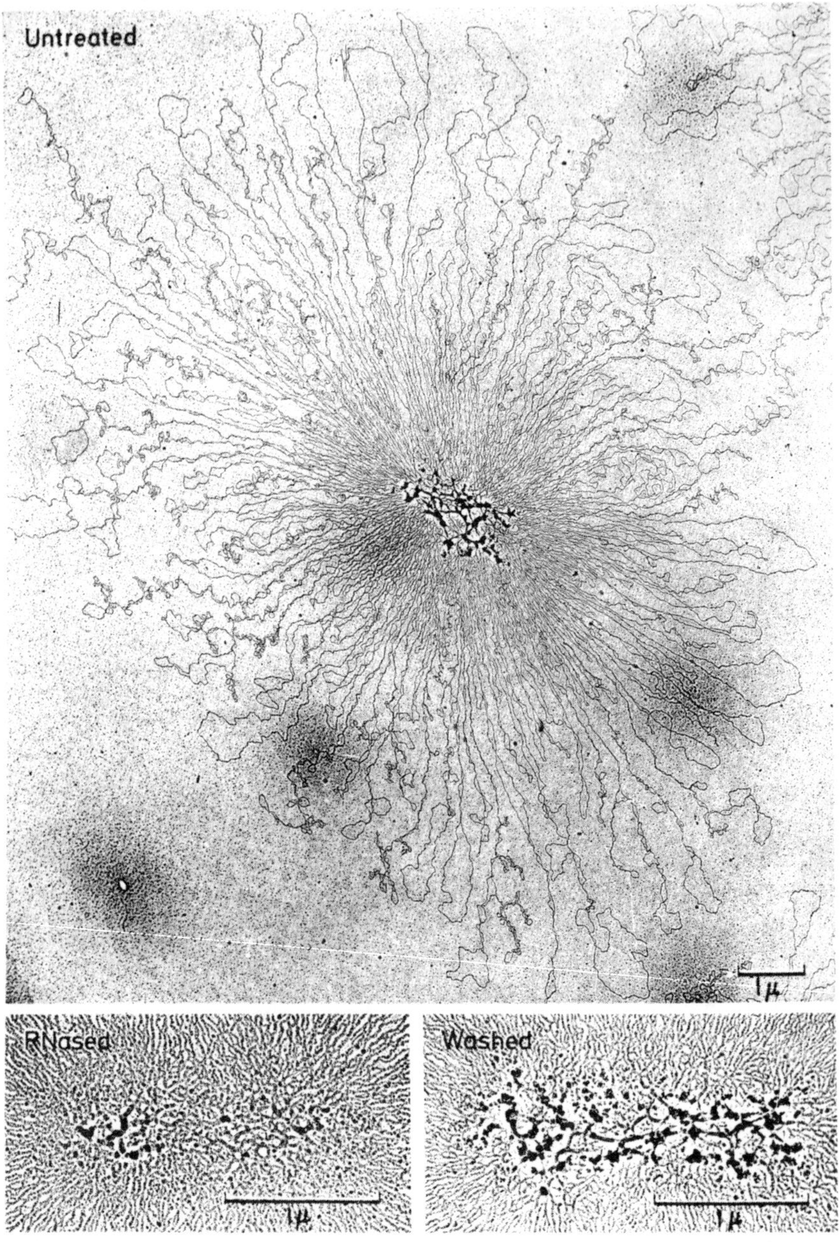

Fig. 4.2 The cell envelope-free nucleoid from *Escherichia coli* spread on 0. 4 M NaCl. Some 141 loops of DNA can be seen. The material at the centre can be partly removed by treatment with RNase (lower left) and is to be compared with the control (lower right) treated with only buffer. (Picture kindly supplied by Dr Ruth Kavenoff)

conditions used for isolating and handling nucleoids. Moreover, careful autoradiographic studies of whole cells of *E.coli* have strongly suggested that RNA synthesis occurs in the cytoplasm and in association with the cytoplasmic membrane rather than in the centre of the nucleoid. It is conceivable that the isolated nucleoid represents an inversion of the situation in the bacterium itself. We must again conclude that, despite our considerable knowledge of the structure of DNA as the classical double helix, and of its mode of function in supplying the triplet code for transcription to messenger RNA, our understanding of the organization of the DNA in the living cell and certainly of its mode of segregation after replication is still at an elementary stage.

The ribosomes

In sections of bacteria the cytoplasm appears to be filled by aggregates of ribosomes, but occasionally a few rather well spaced ribosomes can be seen, both in sections and in freeze-fractured preparations. They are about 20 nm in diameter and are made up of two subunits. In isolated preparations they sediment in the ultracentrifuge at 70 S, and the two disassociated subunits at 30 S and 50 S.

As with the organization of DNA in bacterial cells, more detailed information about the organization of the ribosomes has come from gently bursting the cells and examining the contents under the electron microscope and by hydrodynamic methods, than from attempts to examine whole cells. Ribosomes are an essential part of the translation machinery in cells ensuring that the code which has already been transferred from the DNA to the messenger RNA (mRNA) is then used to assemble and join together amino acids in the correct order to make functional proteins. A high proportion of the ribosomes might, therefore, be expected to be in intimate contact with the mRNA. Examination of the contents of a fragile strain of *E.coli* that could be burst by diluting a culture of exponentially growing cells in water, showed that a very high proportion of the ribosomes were so arranged. Groups of ribosomes of up to forty particles were found attached to the genome by strands of mRNA, thus forming polysomes. The arrangement of the polysomes along the genome was irregular, and the sites of their attachment were marked by the presence of an irregularly shaped granule of about 7.5 nm in diameter. It is possible that these granules were RNA polymerase molecules.

There has been much discussion about a possible separate population of ribosomes attached to the cytoplasmic membrane. Indeed, theories for the export of proteins from the cell demand that some ribosomes should have intimate contact with the membrane. Proof that such a separate population exists is very difficult. When membranes are isolated from bacteria, the extent of their association with ribosomes is controlled, among other factors, by the concentration of magnesium ions in the suspending fluid. Attempts to decide which associations are meaningful and which are non-specific have not progressed very far.

Ribosomes consist of about two-thirds ribonucleic acid and one third protein. Rather complete and precise data are available for the species of proteins and nucleic acids present in both of the sub-units of the most-studied particles from *E.coli*. At the forefront of modern research in this subject are attempts to understand the ways in which these components are organized to give functional ribosomes. When the disassociated 30 S and 50 S subunits are negatively stained and examined with the electron microscope, or when freeze-dried preparations are examined by tungsten shadowing, distinctive shapes are found (see Figs. 4.3 and 4.4). Both

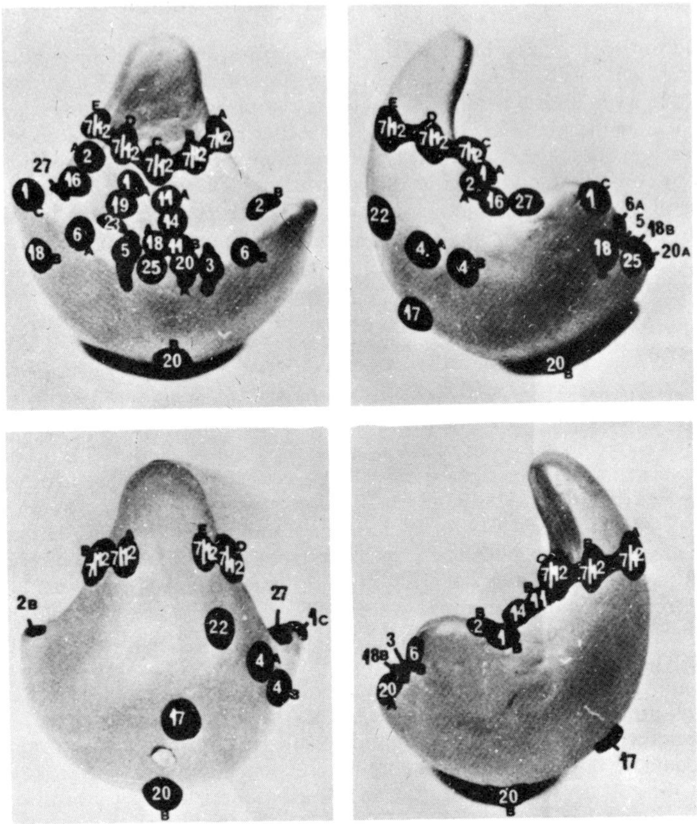

Fig. 4.3 Four views of a three-dimensional model of the 50 S subunit of ribosomes showing antibody binding sites for 19 proteins. (From Brimacombe *et al.*, 1978)

subunits are highly asymmetric and complicated in form; the 30 S subunit can be described as 'embryo-like' whilst the larger 50 S subunit is 'armchair-like'.

Ribosomal proteins Two-dimensional gel electrophoresis has shown the presence of 21 different proteins (S1–S21) in the 30 S subunit and 34 proteins (L1–L34) in the 50 S subunit. All these, except two pairs (L7/L12 and S30/L26), represent quite different proteins having degrees of homology no greater than expected on a random basis. The stoichiometry is such that the subunits contain one of each, except L7/L12 where there are multiple copies. Earlier results which claimed considerable heterogeneity among the populations of ribosomes can probably be attributed to the methods used to make the preparations. The 30 S subunit contains a 16 S RNA whilst the 50 S subunit contains a 23 S RNA species.

The molecular weights of the proteins are rather low, ranging from 5380 daltons for L34 to 23 140 daltons for S4, and the amino acid sequences of a large number of them have been determined. A number of the isolated proteins have been studied by hydrodynamic and small-angle X-ray scattering methods and their shapes have been determined. A considerable proportion are elongated and rather few are globular.

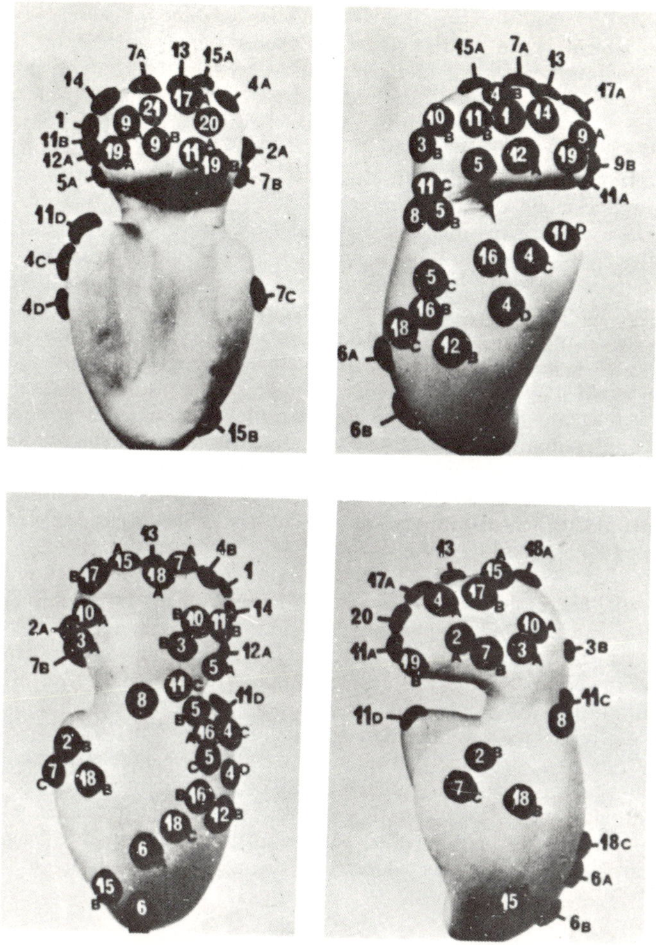

Fig. 4.4 Four views of a three-dimensional model of the 30 S subunit of ribosomes showing antibody binding sites for 21 proteins. (From Brimacombe *et al.*, 1978)

Determination of the arrangement of the numerous protein and nucleic acid components within the subunits cannot be achieved easily by the application of any one set of techniques. Four major methods have so far been applied.

(1) *Immune electron microscopy* This somewhat unfortunate term applies to the attempt to determine the location of the proteins on the ribosome surface by the use of specific antisera which are usually coupled with ferritin to make their subsequent positions on the ribosome particles visible under the electron microscope. The method depends on absolute antigenic specificity of the individual proteins, accessibility of the antigenic sites to purified IgG-antibody, and the unique shape of the subunits. If they were, for example, perfectly spherical, specification of the

absolute position of any given antibody molecule would be very difficult. In Figs. 4.4 and 4.5 the sites at which the antibodies react are labelled. The positions of 21 proteins in the 30 S subunit and most of those in the 50 S have been identified. Six of the proteins in the small subunit, and six on the larger one showed widely scattered multiple binding sites, so these proteins must have highly extended conformations within the subunit. Five other proteins on the 30 S unit had two binding sites separated by 5 to 8 nm and are therefore likely to be only slightly elongated. The remainder showed a single site, but it must always be remembered that only those sites at the surface will be accessible to antibody and able to react. The presence of single binding sites cannot, therefore, be taken as certain evidence that the proteins have a globular or non-extended conformation in the ribosome. Multiple binding sites, on the other hand, for proteins such as L7 and L12 would be expected because more than one copy is present in the 50 S subunit.

(2) *Protein-protein cross-linking* In this method intact ribosomal subunits are treated with bifunctional reagents such as, for example, mercaptobutyrimidate. After reaction of this reagent with amino groups in the protein the mercapto groups are cross-linked by mild oxidation. Depending on whether the cross-linkage can be readily removed the proteins joined together can be identified either by immunological methods, if splitting is difficult, or by gel electrophoresis, if it is not, as in the example of the mercapto reagent mentioned above. Table 4.1 shows the results of protein cross-linking data for the 30 S subunit.

(3) *Neutron-scattering* This method demands the presence of two deuterated proteins in the subunit. Therefore the particle has first to be broken down and then reconstituted *in vitro* with the deuterated proteins included. The distance between the centres of mass can be deduced by neutron-scattering. Such distances between pairs of proteins on the 30 S subunit are shown in Table 4.1. Many of the results agree rather well with those from the immune electrophoresis and the cross-linking methods, although discrepancies have also been found.

(4) *Energy transfer method* This method again involves an *in vitro* reconstitution of subunits and depends upon measuring the efficiency of energy transfer between two incorporated proteins, each labelled with a different fluorescent dye. Most of the 30 S subunit proteins have been labelled and the distances apart have been measured for upwards of fourteen selected pairs of them.

Table 4.1 Protein–protein cross-linking data from the 30 S subunit

S2–S3	S4–S5–S8	S5–S13	S11–S18–S21
S2–S5	S4–S6	S6–S14–S18	S12–S13
S2–S8	S4–S8	S6–S18	S12–S20
S3–S4	S4–S9	S7–S8	S12–S21
S3–S5	S4–S12	S7–S9	S13–S17
S3–S9	S4–S13	S7–S13	S13–S19
S3–S10	S4–S17	S8–S11	S14–S19
S3–S12	S5–S8	S8–S13	S18–S21
S4–S5	S5–S9	S11–S13	

IF2–S1, S2, S11, S12, S13, S14, S19.
IF3–S1, S7, S11, S12, S13, S19, S21.

It would seem likely that concerted attack by these four methods, providing confirmatory and complementary data, will eventually yield a complete picture of the interrelations and topographical distribution of the proteins in both subunits of the ribosomes from *E. coli.*

The ribosomal RNA Compared with the development of our knowledge about the organization of the proteins in ribosomes, our knowledge of the nucleic acids is less satisfactory, even though, or perhaps because, only three species are involved, namely the 5 S, the 16 S and the 23 S RNAs. The ribosome is undoubtedly part of the translation apparatus in all cells, but it is perhaps first and foremost the supervisor of specific interactions between nucleic acids. The specific codon sequences of the mRNA are recognized by the specific anticodon sequences on the tRNA. The polypeptide is generated between the adjacent amino acyl and peptidyl tRNA molecules by peptidyl transferase activity. The ribosomal function may be principally to fix and maintain the tRNAs in their correct position once the codon-anticodon recognition has taken place.

Specific sequences in the 5 S and 16 S RNAs in the ribosome have been proposed to bind with the tRNA and mRNA respectively. It is suggested that the TψC loop of tRNA combines with a complementary sequence in the 5 S RNA, mostly by base pairing. The 3′-terminal sequence of 16 S RNA is thought to be involved in mRNA recognition (the Shine and Dalgarno hypothesis). In support of the latter suggestion an RNA fragment from a protein initiator region of R17 phage RNA and an oligonucleotide fragment from the 3′-terminal end of 16 S RNA have been isolated as a dissociable complex and shown to have a seven-base complementary sequence. The entire idea of the ribosomal RNAs having an active functional role in the ribosome rather than only a structural one is as yet less than ten years old. It is highly probable that new and exciting stories will emerge on this difficult part of protein biosynthesis.

Two other areas of great importance are the conformation of the RNAs and the protein-nucleic acid interactions that occur. Some of the difficulties involved in studying this problem may be illustrated by the fact that more than 20 suggestions for the secondary structure of even 5 S RNA have been proposed. Studies on conformation are further complicated by the observation with the 16 S material that results are influenced by the methods used for its isolation, and that its helical content is likely to be different when it is still in the ribosome rather than in a free isolated state. Interestingly, unfolding of the RNA within the ribosome allows random exchanges of the ribosomal proteins. It has also been shown that combination of isolated RNA with ribosomal proteins alters its conformation. Another approach to determining the secondary structures of the RNAs has been to treat the ribosomes or their subunits with reagents that attack specific base sequences. One such reagent is kethoxal which attacks non-paired guanine residues. After such treatment the precise positions within the RNA sequence that have been attacked can be examined. One of the interesting results from this approach is that the reactions of the RNA in the 30 S subunit differ according to whether the subunit has been separated or is still combined with the 50 S subunit; this suggests that large conformational changes occur upon 70 S formation. Ultraviolet radiation is also being used to generate RNA-RNA cross-links in order to investigate further the complex interactions suspected to occur either between different RNA molecules or within molecules.

The peptidyl transferase centre Attempts to unravel which components of the ribosome are involved in specific stages of the translation process have proved difficult. Among the stages receiving particular attention is the peptidyl transferase step in which the polypeptide chain is extended. This centre carries out two reactions, one that of extending the polypeptide chain, and the other a termination reaction.

The reactions are:

(1) AA_n-NH CH R′CO-tRNA + NH_2CH R″ CO-tRNA″
→AA_n-NHCH R′CO-NHCH R″ CO-tRNA″ + rRNA′

(2) AA_n-NHCH RCO-tRNA + H_2O →AA -NHCHRCOOH + tRNA

Ingenious methods have been used to implicate specific ribosomal proteins in these reactions which are carried out by the 50 S subunit. Three of these methods have been: (a) the use of affinity labels of the type $BrCH_2CO$ Phe-tRNA, (b) the use of antibiotics and their analogues, known to act at the peptidyl transferase step, and (c) the reconstruction of ribosomes with modified individual proteins, after partial disassembly by extraction with salt solutions. The results obtained from the three methods are not in complete accordance. From affinity labelling it would appear that proteins L2, L11, L18 and L27 are important; one analogue of chloramphenicol reacted with L2 and L27 whilst another reacted with L16 and L24. The 'modification-reconstruction' approach with modified L11 showed partial inactivation of the peptidyl transferase but with L24 and L25 almost complete inactivation occurred. Affinity labelling studies strongly implicated the 3′ terminal fragment of the 23 S RNA whilst 'modification-reconstruction' showed that the 5 S RNA was vital for the functioning of the 50 S subunit. In assessing these results a number of complexities of the ribosome must be borne in mind. Firstly, the conformation of one protein can be modified by alterations in another. Therefore, when one protein reacts with a reagent other proteins may be modified, so that they are either more or less reactive. Secondly, the affinity labels may label not only proteins directly concerned with the reaction, but nearby ones as well. Thirdly, it is becoming probable that the peptidyl transferase centre catalyses or modifies other steps in the translation process.

Inclusion bodies

Many prokaryotes, grown under suitable conditions, contain particles, often encapsulated by membrane-like material. However, as has already been pointed out (p. 42), the membranes are usually not of the orthodox bilayer type similar to cytoplasmic membranes. Among this type of inclusion body are polyglucoside or glycogen granules. Their shapes may vary from uneven spherical or oval shapes to rod-like structures of 30 nm made of discs 7 nm each with, in *Oscillatoria rubescens*, a central pore. The granules contain highly branched polyglucose of the glycogen or amylopectin type.

Volutin granules contain linear high molecular weight polyphosphate. They are usually spherical in shape and show very large variations in size depending on the species of organism and the conditions of growth. Polyhydroxybutyrate (PHB) granules are also common and are frequently found in organisms such as bacilli. They can account for as much as 50% of the dry weight of the bacteria. Again the PHB granules are seen to be surrounded by a non-unit membrane about which little is known. The molecular weight of the PHB in the granules has been reported to

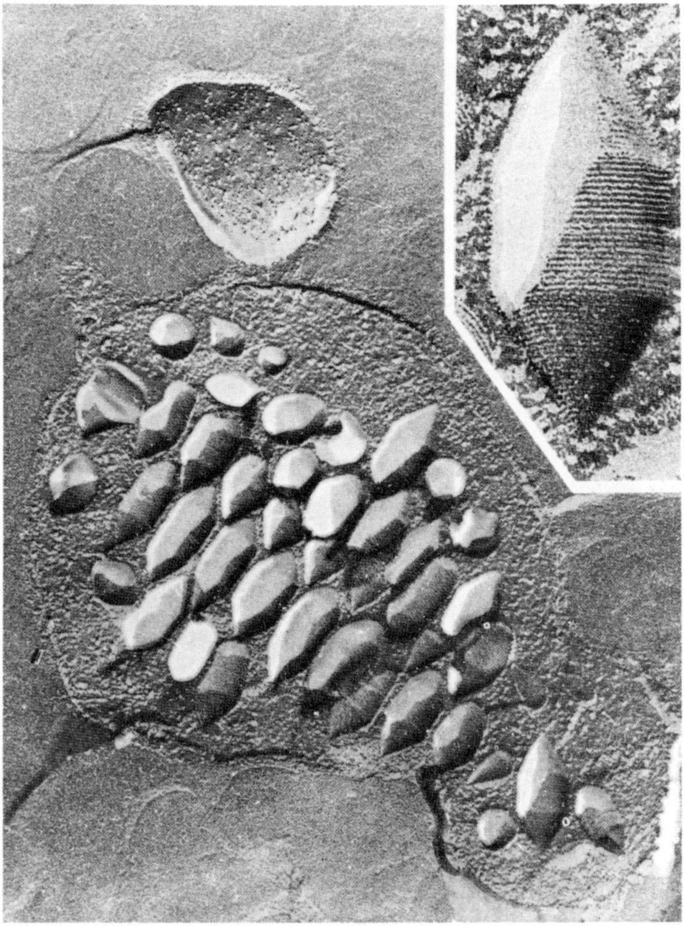

Fig. 4.5 Electron micrograph (magnification × 60,000) showing the gas vacuoles in the cyanobacterium *Prosthecomicrobium pneumaticum*. The inner faces of the gas vacuoles are shown. Inset is a higher magnification (× 210,000) picture of a vacuole, showing the ribs mentioned in the text. (From Walsby, 1978).

vary from 1000 to 256,000 daltons. It would appear that the granules from some organisms such as *B.cereus* may have some internal structure. The presence of an inner core has been described, which during freeze fracturing is plastic and stretches into a horn-like protuberance, surrounded by a layered coat which is in turn contained within a membrane. Such observations suggest that the PHB granules may exist in different kinds of organization or that there are layers of different molecular weight within the granule. The machinery for polymerizing D(-)β-hydroxybutyrate and for degrading the polymer, once formed, is associated with the granule and involves CoA. Activators and inhibitors of the enzyme system reutilizing PHB have been described.

Sulphur globules, a distinctive feature in the Thiorhodaceae, accumulate when hydrogen sulphide is oxidized. The accumulated sulphur is re-oxidized if the supply of hydrogen sulphide becomes limiting. Strictly speaking, these particular sulphur granules are not part of the cytoplasm, because they accumulate in pockets between the wall and the cytoplasmic membrane. In *Thiovulum majus*, *Beggiatoa* and *Chromatium*, on the other hand, sulphur granules accumulate in the cytoplasm proper. They are contained within a monolayer-protein membrane made of subunits of molecular weight 13,500 daltons. It would seem that the sulphur in this type of granule is in a liquid state.

Gas vacuoles occur in many aquatic bacteria where they probably have the very important function, particularly in photosynthesizing micro-organisms, of maintaining the bacteria at a suitable depth in the water. The membranes from vacuoles in cyanobacteria have been subjected to intense study by both chemical and physical means. The subunit protein again has a molecular weight of 14 to 15,000 daltons. A constant feature of the gas vacuoles from all species is that the bounding membrane is organized into ribs 4 to 5 nm wide. Fig. 4.6 shows the structure of a gas vacuole from *Prosthecomicrobium pneumaticum*.

Polyhedral bodies occur in a variety of organisms that use CO_2 as a sole source of carbon, and the name carboxysomes has been suggested for them. The bodies, again surrounded by a non unit membrane, are packed with particles about 10 nm in diameter. In one example these have proved to have ribulose–1, 5-diphosphate carboxylase activity. These particles are among the most interesting of the inclusion bodies in bacteria. Not only have they been shown to have enzymic activity that is crucial for chemolithotrophic bacteria but also those from *Nitrobacter* and thiobacilli contain circular double-stranded DNA of 14 μm in length. If no repetitive sequences are present such DNA could code for 35 proteins. If the carboxysomes from other organisms also prove to contain DNA one could speculate that carboxysomes represent truly sub-cellular bacterial replicating particles. Alternatively they might represent defective bacteriophage particles, although widespread occurrence of similar particles in different genera of bacteria might make this less likely.

Some bacilli form crystalline deposits of which the parasporal crystal of *B. thuringiensis* is particularly well known. It is composed of rod shaped units of a molecular weight of 230,000 daltons, but these can be dissociated into subunits. It is thought that the crystalline structure is related to over-production of the spore coat protein. Various other forms of crystalline material have been described in a range of bacteria. Some such structures have been associated with the presence of bacteriophages.

The cell sap

If bacteria are ruptured and the contents are subjected to prolonged high speed centrifugation the resulting supernatant fluid contains the 'soluble' constituents of the cell. It is a rich source of enzymes. For example, all those concerned with glycolysis of carbohydrates as well as many concerned with their oxidation, hydrolases such as α- or β-glycosidases, amino acid oxidases, proteases and ligases such as those making the precursors of wall polymers, nucleotide phosphorylases and epimerases are found in this soluble fraction. Also found in the so-called cell sap are some cytochromes, ferridoxins and flavodoxins. The organization, if any, of these apparently soluble cell components is quite unknown. One must say 'apparently soluble' because peripheral membrane proteins are rather readily solubilized simply by washing membranes with dilute buffers or chelating agents. It is, therefore, not always clear that some of the 'soluble' cell constituents may not have been associated in the living cell with membranes from which they were removed by the procedures involved in breaking the cells and washing the insoluble constituents.

Summary

Most of our knowledge about both the nuclear material and the cytoplasmic constituents has come from studies of the isolated materials, leaving us relatively ignorant about the arrangement of the components in the cytoplasm of living bacteria. Thin sections of bacteria show less dense areas, often scattered throughout the cytoplasm, which are crossed by thin threads. These represent the nuclear material. Nucleoids have been isolated which consist almost entirely of DNA, RNA and RNA polymerase. The DNA in these is arranged in a series of loops around a more dense central region which consists of RNA. It may be that this arrangement represents an inversion of the situation existing in the living bacteria.

Little information has been gained about the arrangement of ribosomes in bacteria from examination of electron microscope pictures. Isolated material shows ribosomes attached to messenger RNA which in turn is attached to DNA. Our knowledge of the structure of isolated ribosomes is advancing rapidly. Bacterial ribosomes sediment at 70 S and dissociate into two subunits of 50 S and 30 S. Each consists of a large number of defined proteins and 3 species of RNA. By the use of antibodies, protein cross-linking reagents, energy transfer reactions and neutron-scattering, the interrelationships between these proteins in the two subunits is being determined. Work is also progressing on the organization of the RNA. The proteins and nucleic acid species involved in some aspects of the stage of the translation process giving polypeptides are also under study.

Apart from ribosomes a large variety of inclusion bodies can be present in the cytoplasm of many species of bacteria. Many of these have pseudomembranes, usually consisting of a single protein. The molecular weights of the subunits of these proteins range from 13 to 15,000 daltons. The contents of inclusion bodies vary widely from gas in the cyanobacteria to liquid sulphur in organisms such as *Thiovulum majus* and *Chromatium*, and DNA in carboxysomes from *Nitrobacter* and *Thiobacillus* species.

References

BRINACOMBE, R., STOFFLER, G. and WITTMAN, H. G. (1978). Ribosome structure. *Annual Review of Biochemistry* 47: 217–49.

GRUNBERG-MANAGO, M., BUCKINGHAM, R. H., COOPERMAN, B. S. and HERSHEY, J. W. B. (1978). Structure and function of the translation machinery. *Symposium of the Society for General Microbiology* 28: 27–110.

KLEPPE, K., ÖVREBÖ, S. and LOSSIUS, T. (1979). The bacterial nucleoid. *Journal of General Microbiology* 112: 1–13.

KRAYEVSKY, A. A. and KUKHANOVA, M. K. (1979). The peptidyltransferase centre of ribosomes. *Progress in Nucleic Acid Research and Molecular Biology* 23: 1–51.

Bacterial Cell Structure

KURLAND, C. G. (1977). Structure and function of the bacterial ribosome. *Annual Review of Biochemistry* 46: 173–200.

PETTIJOHN, D. E. (1976). Prokaryotic DNA in nucleoid structure. *CRC Critical Reviews in Biochemistry* 4: 175–202.

SHIVELY, J. M. (1974). Inclusion bodies of prokaryotes. *Annual Review of Microbiology* 28: 167–87.

STONINGTON, O. G. and PETTIJOHN, D. E. (1971). The folded genome of *Escherichia coli* isolated in a protein-DNA-RNA complex. *Proceedings of the National Academy of the United States of America* 68: 6–9.

WALSBY, A. E. (1978). The gas vesicles of aquatic prokaryotics. *Symposium of the Society for General Microbiology* 28: 327–57.

5 Bacterial appendages

Flagella

Many bacterial species swim actively by means of filaments attached either to one or both ends of the cells (mono- or bitrichious), or sometimes covering the whole surface (peritrichious). These filaments, known as flagella, are about 12 to 20 nm in diameter and can be 15 to 20 μm long. Some are sheathed whilst others appear not to be. When dried negatively stained preparations are examined by the electron microscope, the filamants appear to describe sine curves with wave lengths of 2 to 3 μm and amplitudes of 0.2 to 0.6 μm. This appearance arises from collapse of the helices in which they originally existed in the living preparation. A variety of mutants with differently shaped flagella have been isolated. These include those with straight flagella which may entirely lack the helical form, or those with curly ones which have a tighter helical structure than in the wild type and wave lengths of 1.1 μm instead of 2.3 μm. The conditions used for the culture of micro-organisms can also affect the form of the flagella. For example, those with a normal wave length can become curly when the pH of the medium is decreased.

There was much early argument about the way in which the flagella filaments are fixed into the bacterial cytoplasm, but analysis of the situation in *Escherichia coli*, salmonellae and *Bacillus subtilis* has now allowed the emergence of a clearer picture for these bacteria. The flagella are made of three parts: (1) the basal body, (2) the hook, and (3) the filament itself (see Fig. 5.1). The basal body consists of a rod about 27nm long bearing, in the structures obtained from *E. coli* and salmonellae, two pairs of discs, and in *B. subtilis* one pair. The diameter of the discs is about 22 nm, whilst that of the rod is 7 to 10 nm. The pair of discs distal from the filament in the Gram negative bacteria is associated with the cytoplasmic membrane and the periplasmic space whilst the proximal pair is associated with the peptidoglycan layer and the outer membrane or lipopolysaccharide. The hook structure is thought to act as a 'universal joint' between the basal body and the filaments.

There is considerable argument in the earlier literature about whether waves passed down the filaments to engender movement of the bacteria, or whether the flagella were rotated like propellers. It was even suggested at one time that flagella might simply be the products of bacterial movement generated by other means, rather than *vice-versa*. The argument has been settled, however, by elegant experiments in which antibodies specific to the flagella were used to tether bacteria to microscope slides, whereupon the bacteria were seen to rotate first in one direction, then in the other. Thus it was clear that flagella function by rotating in a propeller-like fashion.

Bacteria show strong chemotactic behaviour in that they swim towards attractants and away from noxious repellents. They do not, however, swim directly forward in unhesitating straight paths. Recent work on chemotaxis has led to the design of automatic tracking microscopes that allow a precise record of bacterial movement to be obtained. If bacteria are placed in a solution of uniform concentration, they swim forward a short way but then thrash around or *tumble* for a fraction of a second before setting off on another 'run' in a random direction. Addition

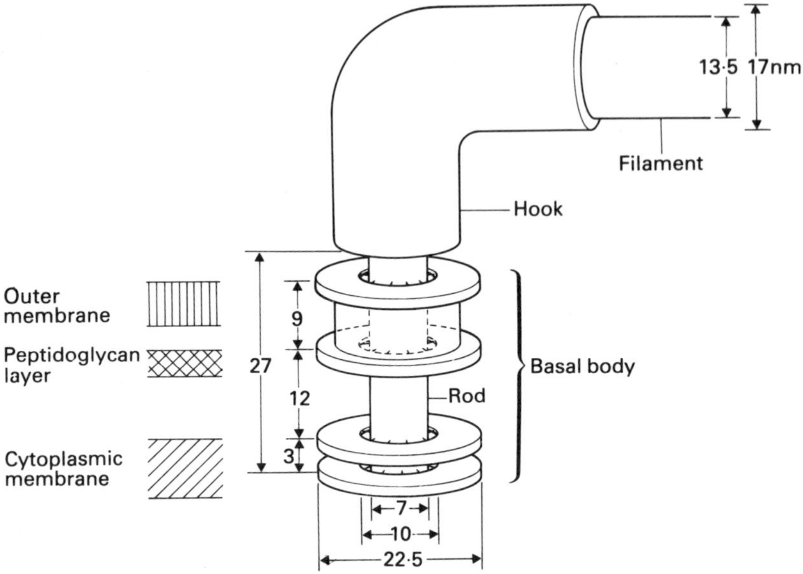

Fig. 5.1 Diagram of the flagellar apparatus of *E. coli*. The figures represent dimensions of the structures in nanometres.

to the solution of an attractant leads to longer runs and less tumbling, whereas addition of repellents increases tumbling. If, as in the experiment already quoted, *E. coli* is tethered to a microscope slide by antibody, addition of attractants causes the cells to rotate counter-clockwise. This would correspond in the untethered organism to a clockwise rotation of the flagella. Addition of repellents has the reverse effect. The tumbling of the cells has also been observed to correlate with a change in the form of the flagella, usually changing from normal to curly. Mutants that do not tumble always rotate counter-clockwise whilst those that almost always tumble mostly rotate clockwise. It is not entirely clear whether a change in rotation of the flagella leads to a change in their form – the latter incidently correlating with a change in the handedness of the helix – or whether the two are independent phenomena.

Chemistry of the flagellar process

The filament The filament is a single subunit protein that varies in molecular weight from 3.3×10^4 daltons to 6.0×10^4 daltons according to the bacterial species. Flagellins from different species differ somewhat in their amino acid composition but cysteine and tryptophan are always absent whilst the amounts of proline, tyrosine and histidine are very low (see Table 5.1). In flagellin from some serotypes of salmonellae but not from others, the unusual amino acid *N*-methyl lysine is found.

Protein subunits of about 4.5 nm in diameter are arranged in hexagonal packing

Table 5.1 Composition of F-pilin, Type 1 pilin and flagellin from *E. coli*

	Molecular proportions of component		
Amino acid	F-pilin	Type 1 pilin	Flagellin
val	21	13	21
ala	15	34	60
gly	15	17	32
ser	11	10	24
lys	10	3	13
leu	9	10	31
met	8	0	2
thr	8	20	41
asp	8	20	64
phe	7	8	5
ile	4	4	18
glu	4	13	40
tyr	2	2	9
trp	2	0	0
his	0	2	1
arg	0	3	10
pro	0	2	4
cys	0	2	0
phosphate	2	–[a]	–[a]
glucose	1	–[a]	–[a]

[a] none reported

in the filaments, leaving a central hole or tube. When we come to consider growth of the filaments the reason for expecting the central cavity sometimes to be filled with material will become apparent. The longitudinal arrangement of the subunits is helical, as shown by optical diffraction and filtering techniques, although claims have been made for a simple longitudinal arrangement in flagella from some species.

Brief exposure of a preparation of flagellar filaments to a wide variety of conditions such as acid, alkali, alcohol, acid-alcohol, buffered osmic acid, formaldehyde, glutaraldehyde, uranyl acetate, freezing and thawing or ultrasonication can break up the filaments into fine wavy fibres. There is an increase in viscosity of the preparations and from the dimensions of the particle the fibres may consist of end to end strings of subunits. Such strings presumably persist because the longitudinal forces between the individual components are greater than the lateral ones. Flagella from some organisms, for example *Vibrio fetus*, behave differently and fibres are not formed.

Exposure to more rigorous conditions of heat or low pH for longer times leads to total disaggregation of the filaments into subunits of flagellin. Conditions necessary for this to happen vary widely according to the source of flagella. For example for those from *Proteus vulgaris* and *B. subtilis* it occurs in less than 5 min at 70°, whereas those from *Clostridium thermosaccharolyticum* withstand 78° for 50 min and those from *Clostridium butyricum* disaggregate rapidly at 58°. Treatment at pH values from 2.0 to 4.0 also leads to flagellar filament disaggregation. Again there are rather wide differences in the pH required to disaggregate material from different species of bacteria. Filaments are also disaggregated by treatment with guanidine hydrobromide, urea, acetamide, alcohols, other polar solvents, both cationic and anionic detergents and sonication.

Flagellin subunits in solution can be made to reassociate to give flagella-like filaments. The conditions under which this happens are rather specific but are as diverse as those for the original process of disaggregation. For example, *Proteus vulgaris* flagellin will not reassemble at all under conditions which are optimal for the protein from *B. pumilis*. Bacillary flagellin will reassemble at temperatures greater than 26° to give either straight filaments (at pH 4.0 to 4.9) or helical normal ones (at pH 5.3 to 6.2). The straight filaments can be converted to a helical form by incubation at the higher pH. Flagellin from salmonellae on the other hand, reassociates at neutral pH in the presence of high salt concentrations such as 33% saturated ammonium sulphate. At low salt concentration the solution must first be seeded with fragments of filaments which then grow by a process seemingly akin to crystallization, but from one end only.

Flagellar filaments also grow from their distal poles when attached to the living organisms, as has been shown by the incorporation of amino acid analogues and of radioactively labelled amino acids followed by autoradiography. When growing cultures of salmonellae are provided with *p*-fluorophenylalanine (PFA), curly flagella are formed. When bacteria grown in broth are transferred to minimal medium containing PFA a proportion of the distal tips of the flagellar filaments assume the curly configuration, suggesting that the abnormal flagellin formed in the presence of PFA has been added at the distal tip. The same conclusion has been drawn from an autoradiographic study of *B. subtilis* using pulses of radioactively labelled amino acids. The question then arises as to whether flagellin is simply excreted by the organisms and added to the tip of the filament from outside, or whether it is passed up the central cavity. Evidence supports the latter possibility. No soluble flagellin can be detected in cultures of bacteria despite the requirement for relatively high concentrations *in vitro* to effect extension of filaments. The rate of extension of flagella decreases exponentially with increasing length, which is most easily explicable by supposing growth of the filament to be proportional to the transit time of the flagellin subunits up the flagellar filament to its tip. Finally, coating the outsides of flagella with specific antibodies does not affect their growth in length.

The hook Connected to the base of the filament is the hook (see Fig. 5.1). It is about 900 nm long and consists of a single protein with a molecular weight of 42,000 daltons in *E. coli* or salmonellae and 33,000 daltons in *B. subtilis*. It is much more difficult to dissociate into subunits than is the filament. Chemical work with this structure was for a long time hampered by the small amount available since it only accounts for about 1% of the weight of the flagellar apparatus. However, mutants are now available that cannot terminate hook formation and form extended polyhook structures. These take the form of curly filaments extending about 1 to 2 μm from the bacterial surface and clearly differ in appearance from the flagellar filaments. They react specifically with antibodies against normal hooks, whereas such antibodies do not cross-react with flagellar filaments. Peptide maps of tryptic digests of hooks and polyhooks are identical. Polyhook material provides larger amounts of hook material for chemical and physical studies.

The basal structure The basal structures of flagella are more complicated than either the hooks or filaments although they again account for only about 1% of the weight of the flagellar apparatus. They have been isolated by using techniques that dissociate the filaments but leave the hooks and basal bodies intact, which can then be separated and purified by differential, isopycnic and velocity gradient centrifugation. Alternatively, mutants are available that form the hook and basal structures but not the filaments. The former can then be isolated by disassociating membrane preparations made from the mutants with detergents. Unlike filaments and hooks, which are each made of only one protein, basal bodies contain at least nine or ten polypeptides, and the number could be as great as thirteen.

The above description is of the components of the flagellar apparatus visible in enterobacteria and *B. subtilis* by the methods available for fixing, staining and examining biological material under the electron microscope. There is, however, genetic and biochemical evidence to suggest that a number of other cytoplasmic and membrane proteins are involved in the total apparatus. If these are to be seen new techniques will presumably have to be applied. It should also be noted that electron microscopic examinations of a wider range of micro-organisms suggests that the picture drawn may not be universal, and that other flagellar structures exist in some organisms.

Genetic control of flagella formation

Three groups of mutants with disturbances either in the formation or the function of flagella have been isolated from *E. coli*, salmonellae and *B. subtilis*. These are designated H or *hag*, *fla* and *mot*. In addition, a wide variety of *che* mutants disturbed in chemotaxis have been obtained. In salmonellae there are two H genes, H1 and and H2, which are structural genes for flagellin formation; in *E. coli* and *B. subtilis* this control is by a single gene, called *hag* in the former organisms and H in the latter. The isolation of mutants was greatly helped by discovery of a bacteriophage called *chi* for which the active flagellar filaments in both *E. coli* and salmonellae act as part of the receptor site. All three genotypes, *hag*, (or H), *mot* and *fla*, are to be found among strains resistant to the phage. However, the majority of the mutants from *E. coli* were originally isolated as strains defective in chemotaxis. All three genotypes were found as well as *che* mutants, some of which are defective in transferring information leading to flagellar activity as a response to chemotactic stimuli rather than being disturbed in the formation of flagella. Mutation of the former genes can lead to one of several phenotypes such as those affected in the efficiency of filament formation, shape of filaments, sensitivity to flagellotropic phage, specificity of flagellar antigens, and flagellar motility. The fla^- phenotype is generally not flagellated, whereas Mot^- is flagellated but non-motile because flagella are paralysed. Apart from these the gene *nml* is present in some serotypes of salmonellae but not in *E. coli*; it controls the methylation of about half the ε-amino groups of lysine in the flagellin. The isolation of *N*-methyllysine from salmonellae flagellin has already been mentioned (p. 70).

Genetic analysis of the very large number of flagellar-motility and chemotaxis mutants available has been carried out in salmonellae by P-22-mediated transduction, by HFr × F^- conjugation, and by colicinogenic factor-mediated transfer. In *E. coli*, the techniques used have been conjugation using F factors and Rec^- hosts, specialized transduction with λ-fla, and P1 mediated transduction. The further

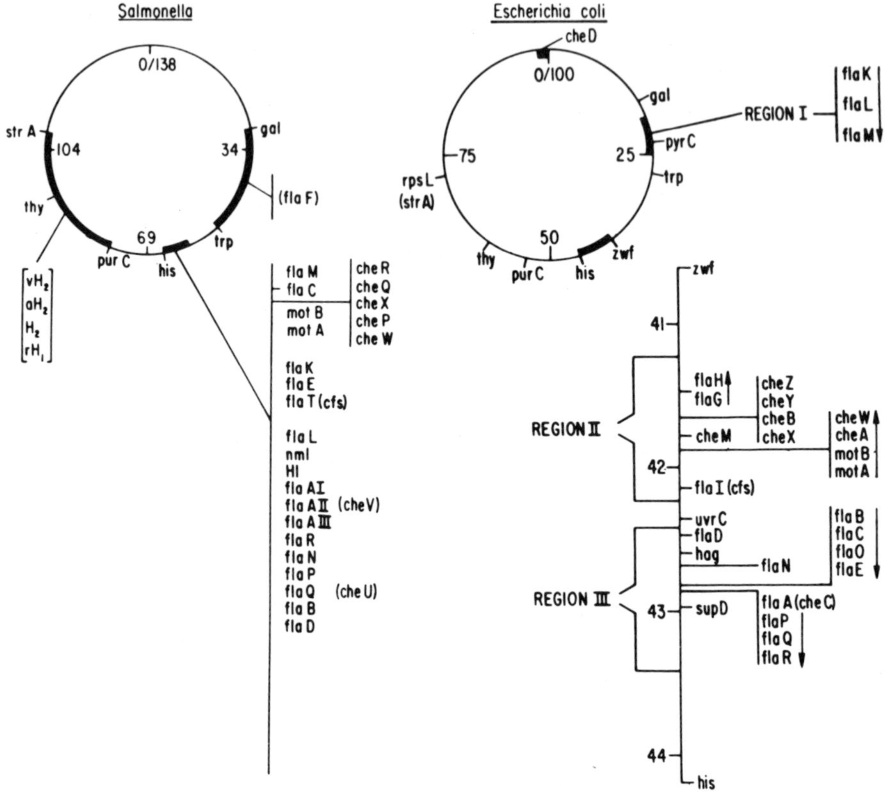

Fig. 5.2 Distribution of flagellar genes on the genetic maps of *Salmonella* and *E. coli*. (From Silverman and Simon, 1977)

organization of the genes in operons in *E. coli* has also been probed by using bacteriophage Mμ to induce polar mutations.

The arrangements of the genes on the chromosomes of *Salmonella* and *E. coli* are shown in Figure 5.2. There are three regions of the *E. coli* genome in which all the flagellar mutations mapped. As can be seen from Figure 5.2, there is a considerable analogy between *E. coli* and the salmonellae in the general grouping of the genes, in particular a large number in both organisms map near the *his* operon. The ordering of the genes, however, is different and whereas H1 in salmonellae has a comparable position to *hag* in *E. coli*, H2 in salmonellae maps further along the chromosome at about 80 minutes compared with 55 minutes for H1; *hag* in *E. coli* maps at about 43 minutes.

Whilst conventional genetic techniques were sufficient for studying the arrangement of the various genes on the chromosomes, examination of gene products involved the use of cloning and recombinant DNA techniques. Dispensable genes were eliminated from bacteriophage λ and replaced by eco R1 restriction endonuclease fragments produced from *E. coli*. The modified phage DNA was then used to infect ultra-violet irradiated *E. coli*. The degree of radiation was such that host protein synthesis no longer occurred, and since the host was a λ-lysogen, the λ-repressor prevented transcription of the λ genome. Therefore the only genes transcribed were those in the new DNA carried on the hybrid phage. Deletions in the hybrid λ generally resulted in the elimination of specific regions of the inserted DNA. It then became possible to correlate the presence of a gene product with a specific genetic complementation activity. The phages could be used to superinfect u.v. irradiated mutant strains defective in the suspected gene. The defective cell was then examined to see whether the particular function had been restored. The gene products could be radioactively labelled and recognized in polycrylamide gels. Specific protein synthesis could not always be obtained, however, using the λ bacteriophage technique. Consequently a second system had to be tried and the colicinogenic factor E1 was chosen as a cloning vehicle. Randomly sheared fragments of the *E. coli* genome were used, and union between vehicle and fragments was achieved by elongating the ends of both with the enzyme terminal transferase, giving short terminal sequences of either polyadenylic or polythymidylic acids. After transformation with the hybrids, a colony bank containing representatives of all regions of the whole *E. coli* chromosome was obtained. Among such strains were those carrying flagellar genes. The hybrid plasmids carrying flagellar genes were then transferred to mini-cells of *E. coli* which have no nucleus but can transcribe and translate plasmid DNA when it is introduced into them. Deletions of regions of the plasmids could be obtained using restriction enzymes and the gene products again correlated with residual complementation activity. A large number of products of the flagellar genes have been recognized and their approximate molecular weight computed (See Figure 5.3).

The problem of assigning function to the various gene products is much more difficult. The *fla* E gene in *E. coli* and the *fla* R gene in salmonellae seem to have the function of regulating the length of the hooks, so that either mutations or deletions of these genes leads to the formation of polyhook-like structures. *Fla* I in *E. coli* seems to have an overall regulatory control over the expression of many of the other flagellar genes. A correlation between some gene products and precursor structures has been claimed as a result of electron microscopic examination of the envelopes of the non-flagellate mutants. The suggested series with some of the genetic loci involved is shown in Figure 5.4.

As has been said, the *hag* locus in *E. coli* controls the formation of flagellin. In salmonellae the situation is more complex and different varieties of flagellar antigen can be recognized, even sometimes within the same clone of bacteria, which is due to the remarkable phenomenon of phase variation. In a diphasic strain, that is one showing phase variation, interactions are occurring between the expression of H1 and H2 which code for different flagellins. Early analysis suggested that H2 can exist in two states, active and inactive. In the active state H1 is repressed and phase-2 flagellin is produced; conversely, when H2 is inactive phase-1 flagellin is formed. The H genes are each activated by a nearby *ah* gene. A repressor gene *rhl* forms a repressor substance for H1. *Ah2* controls the activity of both H2 and *rhl* and may be a promotor for these genes. In *rhl* mutants synthesis of both phase-1 and phase-2 flagellins can occur, and filaments are formed containing both types of flagellin. A number of mechanisms explaining phase variation have been put for-

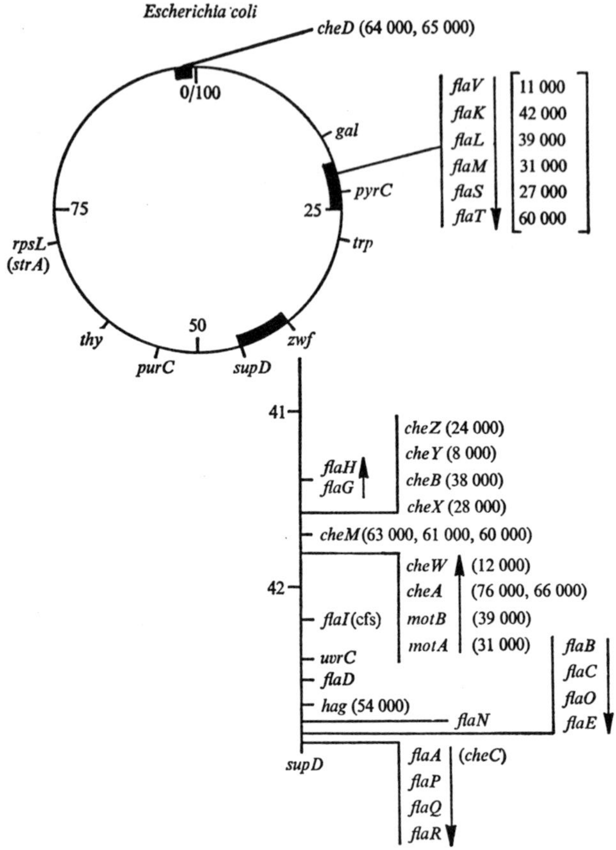

Fig. 5.3 The *E. coli* chromosome with regions of flagellar related genes indicated together with the molecular weights of gene products specified. (From Simon *et al.*, 1978)

ward. Undoubtedly the most successful model has been that which involves the inversion of a segment of DNA adjacent to the H2 gene. Those interested in this fascinating aspect of the molecular biology of flagella formation should read the references given in the review by Silverman and Simon (1977). Formation of flagella in *E. coli* is, perhaps surprisingly, subject to catabolite repression by growth in media containing, for example, glucose. Depression can be achieved by the addition of cAMP and mutants defective in adenylcyclase or the cAMP receptor protein are unable to synthesize flagella. Mutants have also been discovered that synthesize flagella constitutively in the absence of active adenylcyclase and receptor protein under catabolite repressing conditions, as well as at 42° when *E. coli* usually is not flagellate. These mutants allowed a new regulatory site, *cfs*, to be discovered that maps adjacent to *fla* I. The latter had still to be intact for flagella formation to occur.

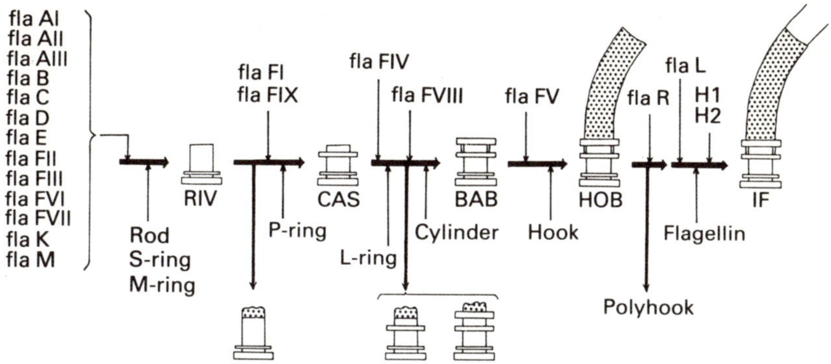

Fig. 5.4 Diagram suggesting some of the functions for the flagellar gene products in flagellar morphogenesis in *Salmonella*. RIV = rod-inner ring complex. M-ring is in the cytoplasmic membrane, S-ring is immediately above it. P-ring is associated with the peptidoglycan, L-ring is associated with the lipopolysaccharide. CAS = RIV-P-ring complex. BAB = basal body. HOB = hook-basal body complex. IF = completed flagellum. (From Iino, 1977).

The flagellar motor

It has by now become quite clear, at least in *E. coli*, salmonellae and *B. subtilis*, that a counter-clockwise rotation of the basal body of flagella applied to a left handed helically arranged filament provides the thrust for the swimming movement of the bacteria. In those with peritrichously arranged flagella the individual filaments form a bundle. Not only will wild type bacteria rotate if their flagella are tethered to a slide by suitable antibody but so will mutants with only a polyhook structure. During the swimming brief clockwise rotations cause periodic tumbling. It would seem likely that the essential rotating part of the rotor is the M-ring of the basal body buried within the cytoplasmic membrane. How does this work? We cannot yet provide a complete answer although certain points have become clearer. The rotor appears to be driven by the proton motive force and not ATP directly, as has been shown by using mutants of *E. coli* and salmonellae blocked in oxidative phosphorylation. Thus the prokaryotes differ from eukaryotic cells which use ATP directly to drive their flagella. Moreover, tethered bacteria can be made to rotate by exposing them to an ionophore such as valinomycin while suspended in a medium of low K^+ concentration. Under these circumstances protons flow into the cell and K^+ flows out.

The proton motive force (Δp) is made up of the charge difference across a membrane, the membrane potential ($\Delta\psi$), and the differential pH across the membrane (ΔpH). It is related to these parameters by the equation:

$$\Delta p = \Delta\psi - \left(\frac{2.3\,RT}{F}\right)\Delta pH$$

where R is the gas constant, T is the absolute temperature and F is the Faraday constant.

In the presence of valinomycin motion ceases as soon as the cytoplasm has been sufficiently acidified by the inflow of protons to lower the ΔpH. Similarly, temporary rotation can be caused by either acidification or alkalinization of the medium; either will do because the sign of the proton motive force is not important. Further examination of this system shows that the angular velocity of rotation is directly proportional to the proton motive force, suggesting that each turn of the rotor requires the flow of a fixed number of protons past the M-ring. It may be significant that the gene products of the *mot* genes, which appear to control rotation of the flagellar motor, are cytoplasmic membrane proteins. Various descriptions exist for discernible regularly arranged structures associated with the M-rings of the basal bodies, but little is yet certain. The threshold value for the proton motive force required to cause rotation of tethered bacteria is very small, being less than 25 mV. That required to cause swimming of untethered bacteria is somewhat greater, being equal to 75 mV, presumably because the work required to move the bacteria laterally through fluids is greater than to rotate them.

Fimbriae and pili

The thin hair-like appendages that project from the surfaces, particularly of Gram negative species of bacteria, were not recognized until electron microscopes were available (Fig. 5.5). Even the best optical systems had insufficient resolving power to allow them to be seen. They mostly range between 3 to 10 nm in diameter, although in a few species they are somewhat thicker. However, they can be up to several μm in length. As soon as electron microscopes became commonly available in the 1940s, fimbriae and pili were recognized in negatively stained or shadowed preparations. There has been some confusion in the literature over nomenclature and these appendages have been referred to as 'filaments', 'bristles', 'fimbriae', 'cilia' and 'pili'. Since it is clear that there are at least two sorts differing in morphology and very different in function, a simple distinction will be adhered to. Non-flagellar appendages not concerned with the transfer of either bacterial or viral DNA will be referred to as fimbriae, whereas those that are so concerned will be called sex pili.

Fimbriae Fimbriae occur commonly among Gram negative bacteria belonging to the Enterobacteriaceae and pseudomonads. Nonetheless, non-flagellar filamentous appendages, apparently not concerned with transfer of DNA, have a much wider distribution and have been reported to occur in *Neisseria*, *Rhizobium*, *Vibrio*, *Morexella* and *Caulobacter*. The only analogous appendages described in Gram positive species have been in strains of *Corynebacterium renale* and *Actinomyces naeslandii*.

In general, fimbriation seems to be more common among freshly isolated strains. The number of fimbriae can be very variable among bacteria even in the same culture, some individuals carrying very few whilst others have a thousand or more associated with their surface. One of the characteristics of all well fimbriated bacteria is the formation of pellicles in static cultures. This is partly, but not wholly, explained by the involvement of fimbriae in adhesion of bacteria both to surfaces and to each other.

Various attempts have been made to classify fimbriae by their morphology and by the properties of the strains that bear them. A well recognized type is referred to as Type 1, common among Enterobacteriaceae. Bacteria that bear Type 1 fimbriae adhere strongly to plant and animal cells including red blood cells. Isolated fimbriae

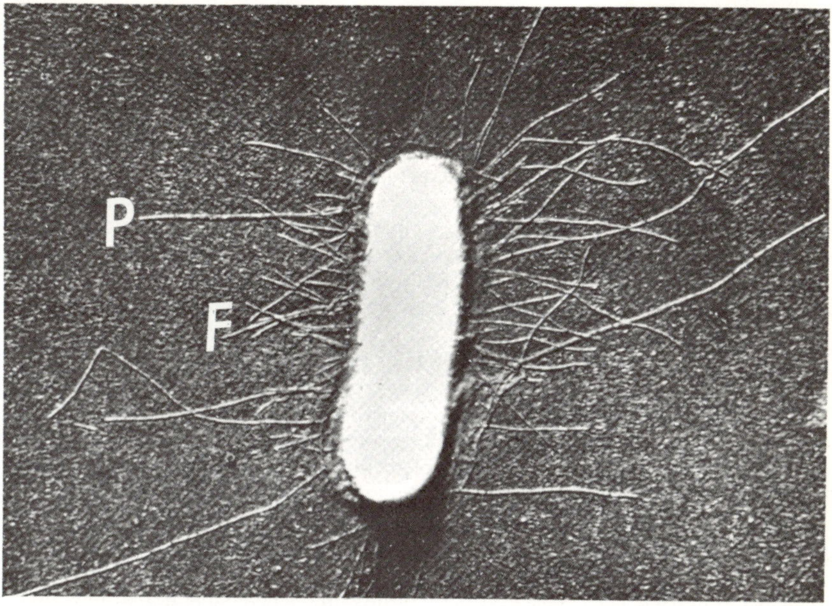

Fig. 5.5 Fimbriae and pili associated with an F + cell of *Escherichia coli*. F = fimbriae, P = pili.

agglutinate the latter, thereby giving rise to the well known haemagglutination reaction. Haemagglutination caused by Type 1 fimbriae must, however, be distinguished from that caused by other antigens such as, for example, K88 which can also appear to be fibrillar in form in dried preparations. Both bacterial adherence and haemagglutination are inhibited by D-mannose and α-D-mannosides. A different type of fimbriae formed by *Klebsiella*, *Serratia marcescens* and *Proteus* cause a strong adherence to non-living surfaces such as glass or cellulose fibres; this is not inhibited by mannosides. This deliberately simplified picture is complicated by finding that different strains of the same species of some bacteria, e.g. of *E.coli*, may have either mannose sensitive (MS) or mannose resistant (MRE) or no adhesins. The MRE adhesins are not associated with visible fimbriae but seem to play an important role in colonization as so-called 'colonization factors'. Some individual strains may have both MS and MRE adhesins.

Isolation and properties Fimbriae can be sheared from bacteria by blending. However, if the bacteria are flagellated as well as bearing both sex pili and fimbriae, methods must be found for separating these appendages. Short periods of blending at lower speeds will remove flagella and sex pili, leaving the fimbriae attached. These can then be removed by more vigorous blending. However, it is clearly safer to start with a *fla*$^-$ strain of *E. coli* or salmonellae which is fimbriated, since then the only appendages present in large numbers will be the fimbriae themselves. After shearing such strains, the whole organisms can be removed by low speed centrifugation, and the fimbriae subsequently deposited at high speed. A purified preparation of fimbriae can then be obtained by centrifugation through a gradient of either sucrose or caesium chloride. Any contaminating outer membrane and other proteins can be removed by treatment with deoxycholate which dissolves the fimriae but leaves insoluble the outer membrane proteins. Subsequent treatment by chromatography in Sepharose B4 in urea leaves pure fimbriae.

Isolated fimbriae consist entirely of subunits of a hydrophobic protein of molecular weight of 16,000 daltons. An analysis of purified isolated fimbriae is shown in Table 5.1. The structures can be dissociated into subunits by reagents such as strong urea and guanidine hydrobromide solutions, that weaken hydrogen bonds without breaking covalent ones. In this they resemble flagella and, as in the latter, the subunits appear to be helically arranged with a central axial hole running down the filament. This is probably about 2 to 2.5 nm in diameter. The appearance of the fimbriae in *Pseudomonas testosteneni* after negative staining led to the suggestion that they might be constructed like F-actin filaments from muscle, from helically arranged strands each consisting of globular subunits.

Genetic control of fimbriation A *fim* gene, subsequently called *pil*, controlling whether or not fimbriae are formed has been recognized in the *E.coli* genome and mapped at 98 min. More recent complementation analysis of a larger number of non-fimbriate mutants has allowed three closely linked cistrons. *pil* A, *pil* B and *pil* C, to be recognized. Over and above this apparently straightforward control of fimbriation, there is the less well understood phenomenon of phase variation, whereby cultures of genotypically fimbriate organisms give rise to phenotypically afimbriate cells. This occurs randomly and is reversible in that phenotypically fimbriate cells always give some non-fimbriate colonies, and *vice-versa*, which makes it a difficult phenomenon to investigate. Worse still, cultures may consist of stably fimbriate cells as well as those that show phase variation. The recognition of a difference in colony morphology between the various strains is, of course, very helpful. The rate of change from phenotypically fimbriate to non-fimbriate bacteria differs from the reverse change, and both are affected by the conditions of cultivation. The genetic basis of the regulation of phenotypic fimbriation is not yet properly understood although recent molecular biological studies with other systems showing similar variation may suggest that *introns* are involved, that is short lengths of DNA that can be excised and reinserted with different polarity. Detailed discussion of this work, however, would take us outside the scope of the present book.

Sex pili The sex pili on F^+ Gram negative bacteria, in particular *E.coli*, which has been extensively studied, can be distinguished from other surface appendages by a number of characteristics; (a) they are somewhat wider than fimbriae being 9 to 10 nm in diameter, (b) they are few in number (1 to 3), (c) they absorb male-specific RNA bacteriophages along their length, and (d) they absorb DNA filamentous bacteriophages by their tips. Several different types of sex pili have been distinguished.

Purified isolated preparations of sex pili appear to consist exclusively, or almost exclusively, of a protein called pilin in which the presence of two atoms of phosphorus and one molecule of carbohydrate, measured as hexose, per mole of protein have been detected. Some doubt now exists, however, as to whether the carbohydrate is covalently attached to the protein. The pili are dissociated into monomers by detergent solutions such as 10^{-3} M sodium dodecyl sulphate but not by reagents that only weaken hydrogen bonds, such as 8 M urea or guanidine hydrochloride, that will disaggregate fimbriae and flagella. The molecular weight of pilin from

E.coli is 11,000 daltons. Cystine, proline, arginine and histidine are all absent whilst valine, alanine, glycine, serine and lysine together account for 70 to 80% of its composition. The protein of sex pili differs from that of Type 1 fimbriae principally by its much lower content of threonine and aspartate. It is thus a very deficient hydrophobic protein. The compositions of F-pilin from Type 1 fimbriae and flagellin are compared in Table 5.1. The arrangement of the subunits which make up the sex pili has not been totally resolved. One suggestion has been that they are rod-like in structure. An axial hole or canal has been described after negative staining, and a dark central line down the centre of pili has been repeatedly described in F-like pili though it is less prominent in I-like ones (see below). More recently examination of preparations of sex pili by X-ray diffraction has allowed the conclusion that each pilus consists of 4 helical arrangements of pilin subunits leaving a hole in the centre 2 nm in diameter. A number of descriptions of a knob at the distal terminus of the pili also exist. The ability of sex pili to absorb DNA phages such as M13 at their tips argues for some specialization of this region, although so far unambiguous chemical identification of a receptor has not been made.

Genetic control and specification of sex pili Sex pili appear to be present whenever plasmids or whole chromosomes can be transferred between bacteria. Thus, apart from the F^+ pili known as F-like pili and indistinguishable structures formed by *col* V bacteria, there are those specified by drug transfer plasmids in f^+R bacteria, colicicin I and Ela plasmids. All sex pili are concerned in some way with the transfer of DNA between bacteria. It is no surprise, therefore, that the genetic control of the formation of sex pili has been investigated by using the large number of DNA transfer deficient, Tra^-, mutants isolated from *E.coli.* Some nineteen *tra* genes on the F-lac plasmid of *E.coli* have been recognized and mapped. Twelve cistrons concerned with pilus formation are present as a single transcriptional operon of 15,000 to 20,000 daltons molecular weight, whilst one, *tra* J, has been shown to have a positive operator-like control over the functioning of the twelve. In contrast, products of two other genes, *fin* O and *fin* P, appear to exert negative regulation. Mutations in cistrons A, L, E, K, B, V, W, C, U, F, H and G all lead to the absence of sex pili, as of course do lesions in *tra* J. Two further recessive pleiotropic regulatory genes, *sfr* A and *sfr* B, map far from the *tra* operon, at 0 and 84 min on the chromosome. On reasonable evidence it has been suggested that *tra* A or cistron 12 is a likely candidate for the structural gene controlling the formation of the protein pilin.

Function of sex pili Despite the recognition some twenty years ago of sex pili as morphological entities concerned with the transfer of DNA between cells, the exact way in which they do so is still a matter for active debate. It is not yet possible to distinguish between two hypotheses, both of which have their adherents. The first is the more obvious of the two, namely that the pili function as tubes with the DNA flowing through their central cavity between the bacteria. However, there are a number of arguments against this simple idea and the alternative is perhaps slightly more popular. This is that the sex pili act as recognition organs and having recognized a recipient cell become adsorbed to it by their tips. They are then thought to retract, dragging the two bacteria together until they have envelope to envelope contact, at which time DNA is transferred between them. There seems no doubt that the sex pili are themselves concerned with DNA transfer, and it does not occur when, for example, they are removed from F^+ *E. coli* cells by blending or mutation. A number of circumstances have been found under which pili retract, e.g. in the presence of cyanide or arsenate in the culture medium. If a male specific DNA phage or pili-specific antibodies are first attached to the pili, this retraction is prevented, and so is transfer of DNA. Moreover, piliated non-retracting mutants have been isolated that are defective in conjugation. However, the whole idea that true mating pairs are involved in DNA transfer has been challenged. It has been pointed out that in a culture in which conjugation is occurring, pairs of cells are relatively rare and tight groups of bacteria are frequently observed. It may be, therefore, that new ideas will have to be found for the mechanism of involvement of sex pili in the transfer of DNA between bacteria.

Capsules

Some bacteria are surrounded by what appears to be a well defined, often wide halo of material. This can be demonstrated by the method of suspending the bacteria in solutions containing colloidal particles in suspension, e.g. Indian ink. The particles cannot penetrate the material of the halo. If the bacterial bodies are stained with a dye such as methylene blue, they often appear under the light microscope as tiny blue dots surrounded by wide transparent areas. Some organisms on the other hand do not form capsules but do form slimes, and the culture supernatants are often highly viscous.

Our knowledge of the chemistry of capsular substances is extensive and goes back to the very early days of microbial biochemistry. One only has to think, for example, of the detailed existing knowledge of the pneumococcal polysaccharides, otherwise known as type substances, providing the immunological specificity of pneumococci, to be convinced.

A large number of polysaccharides have been isolated from capsules and the slime material of bacteria, among them are substances such as the xylans which are of great industrial importance. Others are glucans, levans, and heteropolysaccharides like hyaluronic acid made of *N*-acetylglucosamine and glucuronic acid. Those interested in the chemistry of these substances are referred to books edited by Sutherland (1977) and by Berkeley, Gooday and Ellwood (1979).

Unfortunately it is much easier to say what we do not know about the ultrastructure of capsules, the relation between the bacterial wall and the capsule, or the mechanisms determining whether capsule or slime formation should occur, than it is to give positive accounts of these important aspects of bacterial structure.

A major problem in trying to examine capsular ultrastructure is that capsules are made mostly of water. It is therefore exceedingly difficult to interpret the results obtained after most preparative techniques necessary for electron microscopic examination of specimens. Undoubtedly, differences between the appearance of capsular material around various species of bacteria are seen to exist if, for example, the organisms are examined by the critical point drying method, or by staining with ruthenium red, a dye which combines with acidic macromolecules. In some species, for example, the former technique shows fibrils of characteristic thickness and length. Freeze-etching which should be more satisfactory is somewhat disappointing. Nevertheless, it sometimes shows material that appears to be fibrillar in nature, making up a wide band around the organisms.

The nature of the edge of the capsule is quite unknown. As a dilute sponge of polysaccharide that is being biosynthesized by the growing cell one might expect it to be ill defined. Yet the impression given by Indian ink treated preparations is that in fact it is rather compact. No attempt to examine this aspect of the structure of capsules under the electron microscope has so far given convincing answers.

Interrelationships between capsular polysaccharides and recognized components of the wall are clearly interesting and important. Many capsular substances can be largely removed from bacteria by washing them with solutions of appropriate pH and ionic strength, which may suggest that the capsule is held in place by simple entanglement with the wall polymers. Size, charge and conformation of the capsular polysaccharide would then be of primary importance. On the other hand, washing away the capsule may depend upon the action of enzymes present in low activity which snip a few bonds to liberate the polysaccharides. In some instances this seems possible. For example, some strains of Group C *Streptococcus haemolyticus* have wide well defined capsules, consisting of the heteropolysaccharide hyaluronic acid, which are present only during the exponential phase of growth. If such strains are allowed to grow into the stationary phase for a few hours, they lose their capsules and minute amounts of hyaluronidase can be detected. Other strains have no capsules but form abundant hyaluronidase. This suggests that the capsulated strains may form small amounts of hyaluronidase that hydrolyze enough bonds in the capsular polysaccharide for it to leave the region of the cell surface, and thus account for the disappearance of the capsule.

Summary

Four types of appendages are described: flagella, fimbriae, sex pili and capsules. Flagella enable bacteria to swim by rotating like propellers. They consist of three parts; (a) a helical filament constructed from subunits of a single protein (flagellin) arranged hexagonally, leaving a central hole which runs its length and up which new flagellin units are passed during growth, (b) a hook structure also made of a single protein and (c) a complex basal body consisting of a short rod bearing, in *E. coli* and salmonellae, two pairs of discs, and in *B. subtilis* one pair. One of these discs is buried in the cytoplasmic membrane and its rotation seems to be powered by the flow of protons through the membrane. The formation and function of flagella is controlled by four sets of genes.

The fimbriae are long hair-like structures on the surface of Gram negative bacteria and are involved with the organism's ability to adhere to solid surfaces. They are also constructed from subunits of a single protein. The formation of fimbriae is controlled by three cistrons but is complicated by the phenomenon of phase variation, the basis of which is only now being explored.

The sex pili are wider in diameter than the fimbriae and fewer are formed per cell. They are concerned with the transfer of DNA between bacteria. They absorb male-specific RNA phages along their sides and filamentous DNA phages at their tips. Again they are constructed from subunits of a single protein, but these are probably arranged helically. The protein can be distinguished from fibrillin and flagellin by the presence of a mole of glucose and two moles of phosphate per molecule. Sex pili can contract, which is possibly an essential part of the process of cell-to-cell transmission of DNA. The genetic control of the formation of sex pili is complex and 13 genes have been identified and mapped.

Many bacteria are surrounded by capsules which are commonly of polysaccharide. The chemistry of capsular substances has been studied for many years and the structures of many capsular polysaccharides are known. Little, however, is known about the ultrastructure of capsules or of their relationship to the cell wall and its polymers.

References

ACHTMAN, M. and SKURRAY, R. (1977). A redefinition of the mating phenomenon. In: *Microbial Interactions* pp. 235–69. (Receptors and Recognition, Series B, Volume 3.) Edited by J. L. Reissig. Chapman and Hall, London and New York.

BERKELEY, R. C. W., GOODAY, G. W. and ELLWOOD, D. C. (1979). *Microbial Polysaccharides and Polysaccharases.* (Special Publication of the Society for General Microbiology, Number 3.) Academic Press, London and New York.

DOETSCH, R. N. and SJOBLAD, R. D. (1980). Flagellar structure and function in eubacteria. *Annual Review of Microbiology* 34: 69–108.

DUGUID, J. P. and OLD, D. C. (1980). Adhesive properties of Enterobacteriaceae. In: *Bacterial Adherence* pp. 187–217. (Receptors and Recognition, Series B, Volume 6.) Edited by E. H. Beachey. Chapman and Hall, London and New York.

IINO, T. (1977). Genetics of structure and function of bacterial flagella. *Annual Review of Genetics* 11: 161–82.

OTTOW, J. C. G. (1975). Ecology, physiology and genetics of fimbrae and pili. *Annual Review of Microbiology* 29: 79–108.

ROTH, I. L. (1977). Physical structure of surface carbohydrates. In: *Surface Carbohydrates of the Prokaryotic Cell* pp. 5–26. Edited by I. W. Sutherland. Academic Press, London and New York.

SILVERMAN, M. and SIMON, M. I. (1977). Bacterial flagella. *Annual Review of Microbiology* 31: 397–419.

SIMON, M., SILVERMAN, M., MATSUMARA, P., RIDGWAY, H., KOMEDA Y. and HILMEN, M. (1978). Structure and function of bacterial flagella. *Symposium of the Society for General Microbiology* 28: 271–84.

SUTHERLAND, I. W. (1977). Bacterial exopolysaccharides, their nature and production. In: *Surface Carbohydrates of the Prokaryotic Cell* pp. 27–96. Edited by I. W. Sutherland. Academic Press, London and New York.

TOMOEDA, M., INUZUKA, M. and DATE, T. (1975). Bacterial sex pili. *Progress in Biophysics and Molecular Biology* 30: 23–56.

WILLETTS, N. and SKURRAY, R. (1977). The conjunction system of F-like plasmids. *Annual Review of Genetics* 14: 41–76.

6 Overall Summary

This book describes the structure of common bacteria. Organisms of greater morphological complexity, such as *Rhodomicrobium*, *Caulobacter* and Cyanobacteria are not included, since their underlying structural principles are not particularly different. The three regions of micro-organisms, their envelope layers, their cytoplasm, and their nuclear regions are examined in detail, and reproductions of electron microscope pictures are included to help the reader. A separate chapter describes the known appendages of bacteria, such as flagella, fimbriae, pili, and capsules. Attempts are made to relate these structures as seen by electron-microscope techniques to the chemical structures of the macromolecules isolated from them. In the cell wall, for example, the chemistry of the strength-giving peptidoglycans along with the attached teichoic and teichuronic acids are described in detail.

The arrangement of the membranes in both Gram positive and Gram negative bacteria is described. The status of the internal membranes, the mesosomes, in Gram positive bacteria is discussed. In Gram negative bacteria two distinct layers of membranes exist: the cytoplasmic membrane with appearance and functions similar to the comparable structure in Gram positive organisms and an outer membrane that appears similar under the electron microscope. This latter membrane has a distinct composition and has the function of keeping out large hydrophilic molecules and some hydrophobic substances, while specifically controlling the entrance of substances such as iron and certain vitamins. The proteins involved in the former function are described. Also attached to the outer membrane are the fascinating but complex lipopolysaccharides that extend into the surrounding medium. These are the most important antigens of bacteria such as the Salmonellae that allow them to be serologically typed and recognised by pathologists. The chemistry of the lipid part of bacterial membranes is dealt with inasmuch as it differs from those obtained from plant and animal cells.

In the cytoplasm, the predominant structures seen by the electron microscope are the ribosomes. Work on ribosomal proteins and RNAs is described together with the attempts to understand their arrangement in the two subunits making up the particles. The chemical structure of DNA itself is not considered here since descriptions are to be found in many other books. Attention is rather concentrated on the arrangement of DNA in bacteria and on the attempts that have been made to isolate whole undamaged 'nuclei'. In the chapter on appendages, the chemistry of flagella and fimbriae is described together with the assembly of subunits of the proteins from which they are made to form the helical structures visible under the electron microscope. Attention is also given to the genetic control of the formation of flagella because a rather complete and satisfactory picture of the genes involved and their protein products had now emerged.

Glossary

Acholeplasma: Wall-less micro-organisms having many different shapes. Have some characteristics in common with bacteria. Formerly called *Mycoplasma.*

Actinomycetal: Filamentous prokaryotes with several nuclei per cell. Previously thought to be microfungi.

Antibody: A protein produced by an animal in response to an antigen, reacting specifically with the antigen.

Antigen: A high-molecular weight substance of foreign origin which when introduced into an animal gives rise to the formation of antibody.

Arabogalactan: Polysaccharides made of arabinose and galactose linked to lipids. Present in mycobacterial cell walls.

Archaebacteria: A group of bacteria believed to be of particularly ancient origin, e.g. *Halobacteria.*

Autolysins: Enzymes present in all bacteria that hydrolyse peptidoglycan.

Chemotaxis: The ability of bacteria to swim up a concentration gradient of an attractant or down one of a repellant.

Complementation analysis (Genetic): Introduction of two recessive genes from different parents into the same diploid cell to test whether the genes are complementary with regard to function, wild phenotype results if genes are different.

Conjugation: The passage of DNA from one living Gram-negative bacterium to another, detected by genetic means.

Cyanobacteria: Formerly known as blue-green algae: Complex internal structure but true prokaryotes.

Diploid strain: Each chromosome represented twice, each coming from a different parent.

Eukaryotes: Organisms and cells from organisms in which a nuclear membrane exists containing the chromosomal DNA.

F-particle: Plasmid in male *Escherichia coli* cells enables the chromosome to pass to a female cell.

Feulgen stain: Staining of DNA by fuchsin after brief acid hydrolysis.

Fimbriae: Fine long hair-like projections from bacteria.

Flagella: Spiral, rotating appendages of bacteria that are responsible for their motility.

Forssman antigen: Complex antigen possessed by some bacteria in common with red blood cells and tissues of some animals. In pneumococci it is related to lipoteichoic acid.

Forespore: The body first cut off within the sporangium during sporulation, surrounded by a double membrane.

Freeze-etching: After being freeze-fractured ice is allowed to evaporate from the preparation thus revealing buried surface structures.

Freeze-fracture: Preparation of cells are cooled to $-70°C$ and fractured with a glass or diamond knife whilst under vacuum.

Gram-stain: Differential microscopic stain introduced by Dr Gram. Positive organisms stain violet, Gram-negative ones red.

Inclusion bodies: Small bodies seen by the light-microscope in bacterial cytoplasm.

λ-lysogen: A strain of *Escherichia coli* with λ-bacteriophage-DNA integrated into its chromosome.

L-forms: Derived from bacteria, have no walls and grow under special conditions. May be bacterial mutants with lesions in peptidoglycan synthesis.

Lipid A: Complex lipid of lipolysaccharides, containing phosphorylated *N*-acylated glucosamine. Hydroxy fatty acids present.

Lipomannan: Succinylated and acylated mannose polysaccharide of *Micrococcus luteus.*

Lipopolysaccharide: Complex polysaccharides joined to Lipid A, consisting of a side chain and core region.

Lipoteichoic acid: Polyolphosphates, acylated at one end which is fixed in the cytoplasmic membrane. Extend through the walls of Gram-positive bacteria.

Mesosomes: Internal membranes seen in fixed and stained bacteria. Particularly prominent in Gram-positive specie.

Mycolic acids: Long chain cyclopropane fatty acids. Present in Mycobacteria and related genera.

Negative staining: Organisms are treated with solutions of ammonium molydate or sodium silicotungstate before examination by the electron-microscope. The dark areas seen are depressions filled with stain.

Nonunit membrane: Appears as single layer, 2.0–4.0 nm thick, in thin sections.

Nucleoid: The whole chromosomal DNA of bacteria supposedly isolated in an undamaged state.

O-antigen: Dominant antigens which in Salmonellae are lipopolysaccharides. Antigenic specificity due to carbohydrate side chains of lipopolysaccharide.

Peptidoglycan: The strengthening and shape-maintaining polymer in bacterial walls.

Plasmids: Small circles of DNA bearing genes often capable of behaviour independent of the chromosome.

Plasmolysis: Shrinkage of the cytoplasm of cells away from the wall.

Primordial cell wall: A layer of wall differing from the cortex and present in a germinating spore.

Prokaryotes: Micro-organisms in which no nuclear membrane contains the chromosomal DNA.

Protoplast: The bodies left after the complete enzymic removal of walls from Gram-positive bacterial species. Only stable in strong solutions of low molecular weight substances.

R-layer: The thin peptidoglycan and lipoprotein layer of Gram-negative bacteria.

Recessive mutation: Phenotype only evident in the absence of the wild-type dominant gene. Usually a recessive mutation results in loss of functional product.

Restriction enzymes: Endonucleases hydrolysing DNA at specific bonds defined by the nucleotide sequence.

Rocket immunoelectrophoresis: Technique of two dimensional gel electrophoresis. Antibodies are incorporated in one gel. Name derived from shape of tracks formed.

Sex-pili: Long thin projections from some Gram-negative bacteria behaving as males in conjugation.

Sphaeroplast: Bodies left after enzymic rupture of the peptidoglycan layer in Gram-negative cells. Still have wall components present. Stable only in strong solution of low molecular weight solutes.

Spores: Bodies formed by some bacterial species that are light-refractile, and extremely resistant to heat and chemicals. Will germinate to form normal bacteria.

Sporangium: The cell in which a spore is formed.

Teichoic acid: Polyol-carbohydrate-phosphate polymers linked to the peptidoglycan of bacterial walls. Present only in Gram-positive bacteria.

Teichuronic acid: Acidic wall polysaccharides linked to peptidoglycan. Present in Gram-positive bacteria.

Transduction: The introduction of DNA from one bacterium to another by means of a bacteriophage which has integrated some of the bacterial chromosome.

Urkingdom: A taxonomic term meaning a collection of biological kingdoms (e.g. animal, plant) of which the prokaryotes are one.

Volutin granules: A name given by microscopists to the more prominent intracellular granules in bacteria.

Index

Index